Human Rights in Environmental Perspective

Human Rights in Environmental Perspective

Sarat C. Joshi

AKANSHA PUBLISHING HOUSE
NEW DELHI • GUWAHATI • VISAKHAPATNAM

AKANSHA PUBLISHING HOUSE
4649-B/21, Ansari Road
Darya Ganj, New Delhi - 110 002
Ph.: 23263193 / 9811582579 / 09435116718
Fax: 011-23263193
Email: ektabooks@yahoo.com
www.akanshapublishinghouse.com

Human Rights in Environmental Perspective

First Published 2010
Reprint 2019
ISBN 978-81-8370-216-4

PRINTED IN INDIA

Published by Akansha Publishing House, New Delhi and Printed at S.N. Printers, Delhi.

Contents

Preface

> "A small trouble is like a pebble. Hold it too close to your eye, and it fills the whole world and puts everything out of focus. Hold it at proper viewing distance, and it can be examined and properly classified. Throw it at your feet, and it can be seen in its true setting, just one more tiny bump on the pathway to eternity."
>
> —*Celia Luce*

Nature has bestowed the earth with beautiful landscape, plains, high mountains, lush green forests, water springs and lakes, rivers and rivulets, oceans, minerals, air, climate and all the more innumeral species of fauna and flora i.e. life of which homo sapiens is one itself. Collectively these may be called 'natural environment.' But man's greed and misuse, abuse and overuse of environmental natural resources has posed a catastrophic situation and the entire globe is facing natural disasters of one sort or other. Global warming and climate change are the immediate issues, now all nation-states are pondering over.

Here our main concern is healthy environment as a basic human right.

Amidst the plethora of literature available, the present work is unique wherein all pertinent issues are suitably addressed. A unique blend of synthesis of vital information, the following chapters make core of this study—Introduction; Why Sustainable Development?: Issues and Concerns;

Human Rights and United Nations Environment Programme (UNEP); Environmental Justice and Human Rights; Right to Healthy Environment; Environmental Pollution and Human Rights; Environmental Laws and Human Rights; Appendices; Glossary; and Bibliography etc.

The text is supported by different relevant statutes and judgements. As such this has become a handbook of reliable reference to one and the all concerned—academics, professionals and environmental and human rights activists etc.

With the sole aim for furnishing authoritative information, I have subscribed the views of different scholars in the field of environmental studies and human rights. As such sources of information are secondary i.e. published articles, books, magazines etc. available in print form or digitally. I am deeply beholden to all those scholars as well as the sources for benefiting from their works. My seniors, colleagues and friends have generously helped, inspired and guided me in my academic pursuits. I am grateful to them all.

I am thankful to my publishers for envincing keen interest in the present venture and materialising this in the present shape that too within shortest possible time.

Finally, I will be grateful if the learned readers could send their comments and suggestions for enhancing the quality of information contained in the present volume which will be incorporated in subsequent edition with due acknowledgement.

—SCJ

1

Introduction

"... We squander Health
In search of Wealth,
We scheme and toil and save;
Then squander Wealth
In search of Health,
And all we get's a grave.
We live and boast of what we own
We die and only get a stone."

— *Anonymous*

The earth was bestowed with a clean and salubrious climate by the nature. Misuse, abuse and overuse of natural resources have badly affected the environment. Consequently, the entire globe is facing a threat of climate change and global warming and all life on the earth is at the stake of survival.

If nature has bestowed us with a healthy climate, it is our right to have a healthy environment. Healthy environment is now considered a basic human right.

Now first we will put light on environment and related issues.

What is Environment?

The natural environment, commonly referred to simply

as the environment, encompasses all living and non-living things occurring naturally on Earth or some region thereof. The natural environment is contrasted with the built environment, which comprises the areas and components that are strongly influenced by humans. A geographical area is regarded as a natural environment (with an indefinite article), if the human impact on it is kept under a certain limited level.

Earth science generally recognizes 4 spheres, the lithosphere, the hydrosphere, the atmosphere, and the biosphere as correspondent to rocks, water, air, and life. Some scientiests include, as part of the spheres of the Earth, the cryosphere (corresponding to ice) as a distinct portion of the hydrosphere, as well as the pedosphere (corresponding to soil) as an active and intermixed sphere. Earth science (also known as geoscience, the geosciences or the Earth Sciences), is an all-embracing term for the sciences related to the planet Earth. There are four major disciplines in earth sciences, namely geography, geology, geophysics and geodesy. These major disciplines use physics, chemistry, biology, chronology and mathematics to build a qualitative and quantitative understanding of the principal areas or *spheres* of the Earth system.

The Earth's crust, or Continental crust, is the outermost solid land surface of the planet, is chemically and mechanically different from underlying mantles, and has been generated largely by igneous processes in which magma (molten rock) cools and solidifies to form solid land. Plate tectonics, mountain ranges, volcanoes, and earthquakes are geological phenomena that can be explained in terms of energy transformations in the Earth's crust, and might be thought of as the process by which the earth resurfaces itself. Beneath the Earth's crust lies the mantle which is heated by the radioactive decay of heavy elements. The mantle is not quite solid and consists of magma which is in a state of semi-perpetual convection. This convection process causes the lithospheric plates to move, albeit slowly. The resulting process is known as plate tectonics. Volcanoes result primarily from the melting of subducted crust material. Crust material that is forced into the Asthenosphere melts, and some portion of the melted material becomes light enough to rise to the surface, giving birth to volcanoes!

An ocean is a major body of saline water, and a component of the hydrosphere. Approximately 71% of the Earth's surface (an area of some 361 million square kilometers) is covered by ocean, a continuous body of water that is customarily divided into several principal oceans and smaller seas. More than half of this area is over 3,000 meters (9,800 ft) deep. Average oceanic salinity is around 35 parts per thousand (ppt) (3.5%), and nearly all seawater has a salinity in the range of 30 to 38 ppt. Though generally recognized as several 'separate' oceans, these waters comprise one global, interconnected body of salt water often referred to as the World Ocean or global ocean. This concept of a global ocean as a continuous body of water with relatively free interchange among its parts is of fundamental importance to oceanography.

The major oceanic divisions are defined in part by the continents, various archipelagos, and other criteria: these divisions are (in descending order of size) the Pacific Ocean, the Atlantic Ocean, the Indian Ocean, the Southern Ocean (which is sometimes subsumed as the southern portions of the Pacific, Atlantic, and Indian Oceans), and the Arctic Ocean (which is sometimes considered a sea of the Atlantic). The Pacific and Atlantic may be further subdivided by the equator into northerly and southerly portions. Smaller regions of the oceans are called seas, gulfs, bays and other names. There are also salt lakes, which are smaller bodies of landlocked saltwater that are not interconnected with the World Ocean. Two notable examples of salt lakes are the Aral Sea and the Great Salt Lake.

Rivers and Lakes

A river is a natural watercourse, usually freshwater, flowing toward an ocean, a lake, a sea or another river. In a few cases, a river simply flows into the ground or dries up completely before reaching another body of water. Small rivers may also be called by several other names, including stream, creek, brook, rivulet, and rill; there is no general rule that defines what can be called a river. Many names for small rivers are specific to geographic location; one example is *Burn* in Scotland and North-east England. Sometimes a river is said to be larger than a creek, but this is not always the case, because of vagueness in the language. A river is part of the hydrological cycle. Water within a river is generally collected from precipitation through surface runoff, groundwater recharge, springs, and the release of stored water in natural ice and snowpacks (i.e., from glaciers).

A lake (from Latin *lacus*) is a terrain feature (or physical feature), a body of liquid on the surface of a world that is localized to the bottom of basin (another type of landform or terrain feature; that is, it is not global) and moves slowly if it moves at all. On Earth, a body of water is considered a lake when it is inland, not part of the ocean, is larger and deeper than a pond, and is fed by a river. The only world other than Earth known to harbor lakes is Titan, Saturn's largest moon, which has lakes of ethane, most likely mixed with methane. It is not known if Titan's lakes are fed by rivers, though Titan's surface is carved by numerous river beds. Natural lakes on Earth are generally found in mountainous areas, rift zones, and areas with ongoing or recent glaciation. Other lakes are found in endorheic basins or along the courses of mature rivers. In some parts of the world, there are many lakes because of chaotic drainage patterns left over from the last Ice Age. All lakes are temporary over geologic time scales, as they will slowly fill in with sediments or spill out of the basin containing them.

Atmosphere and Climate

The atmosphere of the Earth serves as a key factor in sustaining the planetary ecosystem. The thin layer of gases that envelops the Earth is held in place by the planet's gravity. Dry air consists of 78% nitrogen, 21% oxygen, 1% argon and other inert gases, carbon dioxide, etc.; but air also contains a variable amount of water vapor. The atmospheric pressure declines steadily with altitude, and has a scale height of about 8 kilometres at the Earth's surface: the height at which the atmospheric pressure has declined by a factor of e (a mathematical constant equal to 2.71...). The ozone layer of the Earth's atmosphere plays an important role in depleting the amount of ultraviolet (UV) radiation that reaches the surface. As DNA is readily damaged by UV light, this serves to protect life at the surface.

Effects of global warming: The potential dangers of global warming are being increasingly studied by a wide global consortium of scientists, who are increasingly concerned about the potential long-term effects of global warming on our natural environment and on the planet. Of particular concern is how climate change and global warming caused by anthropogenic, or human-made releases of greenhouse gases, most notably carbon dioxide, can act interactively, and have adverse effects upon the planet, its natural environment and humans' existence. Efforts have been increasingly focused on the mitigation of greenhouse gases that are causing climatic changes, on developing adaptative strategies to global warming, to assist humans, animal and plant species, ecosystems, regions and nations in adjusting to the effects of global warming.

Some examples of recent collaboration to address climate change and global warming are:

(1) The Kyoto Protocol, which is the protocol to the international Framework Convention on Climate Change treaty, again with the objective of reducing greenhouse gases in an effort to prevent anthropogenic climate change.

(2) The United Nations Framework Convention Treaty and convention on Climate Change, to stabilize greenhouse gas concentrations in the atmosphere at a level that would prevent dangerous anthropogenic interference with the climate system.

(3) The Western Climate Initiative, to identify, evaluate, and implement collective and cooperative ways to reduce greenhouse gases in the region, focusing on a market-based cap-and-trade system.

A significantly profound challenge is to identify the natural environmental dynamics in contrast to environmental changes not within natural variances. A common solution is to adapt a static view neglecting natural variances to exist. Methodologically, this view could be defended when looking at processes which change slowly and short time series, while the problem arrives when fast processes turns essential in the object of the study. Although there is no universal agreement on the definition of life, scientists generally accept that the biological manifestation of life is characterized by organization, metabolism, growth, adaptation, response to stimuli and reproduction. Life may also be said to be simply the characteristic state of organisms. Properties common to terrestrial organisms (plants, animals, fungi, protists, archaea and bacteria) are that they are cellular, carbon-and-water-based with complex organization, having a metabolism, a capacity to grow, respond to stimuli, and reproduce. An entity with these properties is generally considered life. However, not every definition of life considers all of these properties to be essential. Human-made analogs of life may also be considered to be life.

Ecosystems

An ecosystem is a natural unit consisting of all plants, animals and micro-organisms (biotic factors) in an area functioning together with all of the non-living physical (abiotic) factors of the environment. Central to the ecosystem concept is the idea that living organisms are continually engaged in a highly interrelated set of relationships with every other element constituting the environment in which they exist. Eugene Odum, one of the founders of the science of ecology, stated: "Any unit that includes all of the organisms (ie: the "community") in a given area interacting with the physical environment so that a flow of energy leads to clearly defined trophic structure, biotic diversity, and material cycles (ie: exchange of materials between living and nonliving parts) within the system is an ecosystem." The human ecosystem concept is then grounded in the deconstruction of the human/nature dichotomy, and the emergent premise that all species are ecologically integrated with each other, as well as with the abiotic constituents of their biotope.

A greater degree of species or biological diversity - popularly referred to as Biodiversity - of an ecosystem may contribute to greater resilience of an ecosystem, because there are more species present at a location to respond to change and thus "absorb" or reduce its effects. This reduces the effect before the ecosystem's structure is fundamentally changed to a different state. This is not universally the case and there is no proven relationship between the species diversity of an ecosystem and its ability to provide goods and services on a sustainable level: Humid tropical forests produce very few goods and direct services and are extremely vulnerable to change, while many temperate forests readily grow back to their previous state of development within a lifetime after felling or a forest fire. Some grasslands have been sustainably exploited for thousands of years (Mongolia, Africa, European peat and mooreland communities). The term ecosystem can also pertain to human-made environments, such as human ecosystems and human-influenced ecosystems, and can describe any situation where there is relationship between living organisms and their environment. Fewer areas on the surface of the earth today exist free from human contact, although some genuine wilderness areas continue to exist without any forms of human intervention.

Biomes are terminologically similar to the concept of ecosystems, and are climatically and geographically defined areas of ecologically similar climatic conditions such as communities of plants, animals, and soil organisms, often referred to *as* ecosystems. Biomes are defined on the basis of factors such as plant structures (such as trees, shrubs, and grasses), leaf types (such as broadleaf and needleleaf), plant spacing (forest, woodland, savanna), and climate. Unlike ecozones, biomes are not defined by genetic, taxonomic, or historical similarities. Biomes are often identified with particular patterns of ecological succession and climax vegetation.

Main Challenges

Remarkably, it is the common understanding of *natural environment* that underlies environmentalism — a broad political, social, and philosophical movement that advocates various actions and policies in the interest of protecting what nature remains in the natural environment, or restoring or expanding the role of nature in this environment. While true wilderness is increasingly rare, *wild* nature (e.g., unmanaged forests, uncultivated grasslands, wildlife, wildflowers) can be found in many locations previously inhabited by humans.

Goals commonly expressed by environmental scientists include:

(1) Eduction and clean up of pollution, with future goals of zero pollution;

(2) Cleanly converting non-recyclable materials into energy through direct combustion or after conversion into secondary fuels;

(3) Reducing societal consumption of non-renewable fuels;

(4) Development of alternative, green, low-carbon or renewable energy sources;

(5) Conservation and sustainable use of scarce resources such as water, land, and air;

(6) Protection of representative or unique or pristine ecosystems;

(7) Preservation of threatened and endangered species extinction;

The establishment of nature and biosphere reserves under various types of protection; and, most generally, the protection of biodiversity and ecosystems upon which all human and other life on earth depends.

Very large development projects - megaprojects - pose special challenges and risks to the natural environment. Major dams and power plants are cases in point. The challenge to the environment from such projects is growing because more and bigger megaprojects are being built, in developed and developing nations alike.

It is pertinent to note that the protection and conservation of environment protection since centuries is a worldwide issue. Environmental crisis has propagated a movement which is undertaken by scientists, social workers, and political leaders even International Organisations. In this regard role of International Cooperation has also assumed tremendous significance. World political leaders took part in International Conferences with a sincerity to curb down environmental problems of their own countries. The severity of the environmental problem has now been proved a major threat to human life. Meaning of protection of life under various national and international documents was only eradication of poverty but later years it also encompasses the clean environment for human being. Rapid degradation of environment by industrial development which were in fact opted to achieve economic flourishment for poverty eradication has caused a great problem to existence of human life. Clean environment has now become essential to human life. Therefore, on account of multidimensional effect of pollution, right to healthy environment is universally accepted as a basic human right universally.

Concrete concern towards rational use and conservation of natural resources of the biosphere was first of all raised in 1968 from the platform of the United Nations, Educational, Scientific and Cultural Organisation (UNESCO), held a special inter-governmental conference where experts had determined the scientific principles of this rational problem. The results of this conference are closely related to a number of important proposals on the environment which arc being given high level consideration by International Organisations.

Later on an international conference held in Stockholm, Sweden in 1972. One of the result of this conference was the adoption of a Declaration on the Human Environment. In a major inter-governmental conference on environment programme (UNEP) in 1977, the Executive Director warned that 'Environmental Education is now a matter of life and death". In 1981, African Charter on Human and Peoples Right was promulgated of which Article 24 lays down that:

> "All people shall have the right to general satisfactory environment favourable to their development."

It is remarkable to note that the African Charter was first International instrument of Human Rights Law. It declares 'right-to environment' as a human right and emphasizes the importance of environment being "general satisfaction' and favourable to ... development."

Further, the United Nations Conference on Environment and Development (Earth Summit) at Rio de Jenerio (Brazil) in June, 1992 minutely reached to the observations that, every citizen has the right to live in environment worthy of human existence. This is natural as the concept of Human Rights is also expanded from its social ambit. Prior to the realisation of environmental degradation human rights were interpreted in terms of social development. But later on when the sporadic industrial development started rapid environmental pollution affects the survival of life. Thus, development and polluted environment both have upset the rights to life and health in other aspects.

Obviously human rights have to be interpreted in terms of healthy environment. Rio-declaration proclaims that: "Human beings are the centre of concerns for sustainable development, They are entitled to a healthy and productive life in harmony with nature."

India took part a leading role in Stockholm Conference on Human Environment in 1972. As of new very recently in Sec. 2009 India has declared its commitment to cut off 20% carbon emission though, its progress may hamper adversely. This shows India's realisation of deadly portens of pollution and commitment to eradicate the pollution for well being of the people. Such conference held under the aegis of United Nations to establish the natural environment as a basic right The right to a healthy environment is natural, inalienable and fundamental right of the people. Therefore, the Declaration acts as an eye opener to mankind and the people in the world. On the essential and imperative need to protect the environment Declaration can be viewed from any direction and finally prove an important milestone for International environmental movement. The proclamation felt that protection of environment is an important need of the time.

It is recapitulated in a part as cited below:

> "Man is both creature and moulder of his environment which gives him physical sustenance and affords him the opportunity for intellectual, moral, social and spiritual growth. In the long and tortuous evolution of the human race on this planet a stage has been reached when through the rapid acceleration of science, and technology, man has acquired the power to transform his environment in countless ways and on an unprecedented scale. Both aspects of man's environment, the natural and the man made, are

> essential, to his well-being and to the enjoyment of the basic human rights. Even the right to life itself."

Remarkably, the United Nations proclamation also adopted that all governments of the world must realise human environment a major issue. Appropriate actions must be taken by these governments in this regard. In a better environment more needs and hopes of human can be achieved. Apart from governmental initiatives man himself must be committed to build a better environment by using his knowledge in collaboration with nature.

There should be a goal of whole mankind to defend and improve the human environment for present and future generations. United Nations conference of 1972, calls upon the mankind of a goal of protection of environment to be pursued together with, and in harmony with the established and fundamental goals of peace and world wide-economic and social development At the same time proclamation demands an infliction of responsibility to citizens, communities, enterprises and institutions at every level on achieving the said goal. The conference has burdened upon the governments and people to exert common efforts for the preservation and improvement of the human environment for the benefit of all the people and for their posterity and well-being.

The concerted action plan of this U.N. Conference included three major components, viz. detailed environmental assessment so as to make rational judgments about priorities for action, organised environmental management and supportive action in the form of laws, manpower institutions, and finances to implement the decisions so reached. Since early 1970's man has become aware of the facts, that the rapidly growing conditions of stress on his environment threaten not only his welfare but even his survival, Man's ambition of everlasting enjoyment and comfort had led him to exploit natural resources to the extent of reducing its natural capacities for self stabilisation.

As a consequence of the outright disregard of the impact of the these activities, numerous environmental problems have arisen which seem to be the bare of modern scientific thought and technological advancements. It is only recently that man has come to realise that the environment is an unbroken totality which includes everything that affects the air, water, surroundings and other factors that affect the quality of life on earth planet.

National Concern for India

But it must be remembered that mere declaration of environment a human right at the floor of United Nations was not enough. In consonance the nations have to take some objective steps in this larger issue of mankind. India in 1976, have brought many constitutional amendments in one go, through which environment as a whole is placed at the position of human right.

Originally the Constitution of India did not contain any provisions having a direct bearing on environment. However, some of the provisions do make tangential reference to the problem like Articles 47, 48 and 49 contain directives to government. India was one of the few countries that has provided constitutional safeguards to preservation of the environment. Articles 48-A, 51-A(g) and entries in three Legislative lists are inserted in the Constitution. These provisions contain a clear mandate to the citizens to protect and improve the environment. Therefore, they constitute the basic policy framework for all the legislative and executive programmes in this regard.

In fact, the first and foremost amendment of the

Constitution with regard to the environment has paved the way to do liberal and socialists attitude towards environmental development. In the preamble of the Constitution for the words "Sovereign Democratic Republic" the words "Sovereign Socialists Secular and Democratic Republic" were substituted. The importance of this amendment can easily be visualized. Though India has always been following the ideal of socialist pattern of society, yet it was for the first time this ideal was given a constitutional recognition by the 42nd Constitutional Amendment Act 1976, The problem of environmental pollution is a social problem affecting as it does the society at large socialistic pattern of the societies imposes a number of new obligations on the state.

Inversely state have to eradicate social hazards and social evils and to ensure social justice. The environmental pollution is one of the most important social problems that a nation is called upon to face. Part II of the Indian Constitution deals with the Directive Principles of State Policies. Art 47 says that the state shall regard the raising of the level of nutrition and the standard of living of its people and the improvements of public health as among its primary duties.... This article provides for the improvement of the public health as one of the primary duties of the state and it could very well include by implication of the protection and improvement of the environment, for the control and prevention of environmental pollution necessarily and primarily comes within the concept of the improvements of the public health. New insertion of Article 48-A as 'the State shall endeavour to protect and improve the environment and to safeguard the forests and the wild life of the country," has directly intervened to casts an obligatory duty on the state to endeavour to protect and improve the environment. To make it effective few insertions are made in the legislative lists enshrined in the seventh schedule of the Constitution.

Further, addition of new chapter, IV-A of fundamental duties in the Constitution has proved deep concern of the state to protect and preserve the natural environment. Article 51A(g) enjoins that "it shall be duty of every citizen of India... to protect and improve the natural environment including forests, lakes, rivers and wild life, and to have compassion for living creatures."

The principles enunciated commands to live together based on an environmental law to violate which is to gamble with civilized life itself. Addition of an exclusive state directive in the voluminous list of directive principles would be meaningless unless the legislative steps are not taken in conformity with it.

As such the policy goals have been laid down in directives and fundamental duties through 42nd amendment Act, 1976. The inclusion of Article 48-A, 51-A (g) entries to legislative list and the word 'Socialist' in preamble have widened the concept and meaning of the 'life' as enshrined in Article 21 of the Constitution. In absence of healthy and clean environment existence of life is not possible. Even political, social and economic rights are questioned in presence of polluted environment. On increasing pollution hazards environment of man is grossly polluted.

Healthy environment is going to become scarce in the regime of increasing industrial, agricultural, chemical and domestic activities of pollution. This menace is found abundantly in the third world countries, On such account the right to healthy environment has been recognised as a "third generation human rights'. Particularly, in India with the innovative legislative activist as well as judicial activist approach it can be manifestly said that the expansionist meaning of Article 21 can lead us to draw the conclusion that deprivation of right to human life is not the only activity remains to be accomplished but to vigorously enforce the legislations enacted under the national policy for the protection of ecology and environment.

Statutory Efforts

In accordance with National Policy for Environment state has promulgated environmental laws to deal environment directly. These are the Water (Prevention and Control of Pollution) Act 1974, The Air (Prevention and Control of Pollution) Act 1981, and the Environment (Protection) Act 1986. Prior to these laws existing local provisions were neither adequate nor satisfactorily dealt increasing tendency of pollution. Therefore, comprehensive legislations have been laid down to establish unitary agencies in the centre and states in India.

The basic objective of Water Act, 1974 is to prevent, abate and control the pollution of rivers and streams for maintaining or restoring wholesome of such water courses. With regard to preservation of the quality of air and control of air pollution the Air Act, 1981 is promulgated. An integrated approach for tackling the environmental problems relating to pollution is necessary. Therefore, the Boards (Central and States) are constituted by the Acts and deputed to perform the functions assigned under the Acts.

Lately, in 1986 Environment (Protection) Act was passed. Object of this act is to deal environmental problems effectively. Although, there are existing laws dealing directly or indirectly, with several environmental matters, it is necessary to have a general legislation for environmental protection. Existing laws generally focus on specific types of pollution or on specific categories of hazardous substances. Some major areas of environmental hazards are not covered.

As such, in view of reflecting total considerations taken

in Stockholm in 1972, India has promulgated a comprehensive legislation to protect and improve the environment. It is therefore, promulgated to confer decision taken at conference of human environment relating to the prevention of hazards to human beings other living creatures, plants and property This Environment Act of 1986 is comparatively proved better in approach than the other laws enacted for prevention and control of specific factors.

Direct environmental legislations have promulgated in conformity of the directive principles. The fundamental duties must also be mandatory, state should not think its endeavours have come to an end when it has acted on its part i.e. by enacting laws. State as well as citizens both have obligations to protect and promote natural environment in order to have right to environment vis-a-vis right to life.

But, despite all said constitutional and legislative provisions rape of the earth, brutal exploitation of our forests, indiscriminate quarrying, and pollution of our water and air is unabated. It is in India only that one of the wonders of the world the Taj Mahal at Agra is suffering stone cancer due to the damage caused by neighbouring 510 industries. Millions of people on the banks of the river Ganga are heading towards an environmental catastrophe. The water of Ganga at Kanpur consisted 29.200 units mg/ml of iron when the maximum limit is 0.3 units, and manganese was found to be 0.900 mg/ml when the WHO limit is 0.5 mg/ml. Every six months more topsoil gets washed away than has been used to build all the bricks houses across the country. Seventy percent of all the available water in India is polluted. About 73 millions workdays are lost due to water related diseases. Cherrapunjee is no more wettest spot on the earth. Delhi is the most polluted city in the country where suspended particulate matter (SIIM) in the air rose alarmingly despite various steps taken to protect the environment. River Yamuna smells like a dirty drain in Delhi.

Judicial activism has done some service to mankind of late. Judiciary through various cases has illustrated the increasing importance of private rights of action in environmental litigation. Judicial attitude in this direction is worth considering because this is a question of enforcement of right to life. A private right of action might better be termed a "citizen enforcement action.' Promoting the purpose of legislative policy through citizen initiative is an important policy goal behind a private right of action may be defined a right under a substantive statute as a private party's action to enforce a regulatory standard. The judiciary has exhibited an enlightened judicial creativity and foresight whenever it had an opportunity, to decide issues relating to environment. The judiciary has been very active in taking up the cause of the environment.

Judicial verdicts have brushed out the technical hindrances by devising new tools of legal interpretation and construction. It is the judiciary by giving directives to the government have accelerated the environmental considerations. In *M.C. Mehta V. Union of India*, following observations' are made.

> "Having regard to the grave consequences of the pollution of water and the need for protecting and improving the natural environment which is considered to be one of the fundamental duties under the constitution (vide clause (g) of Art. 51-A of die constitution) we are of the view that it is the duty of the Central Govt. to direct all the educational institutions throughout India to teach atleast for one hour in a week lessons relating to the protection and

> the improvement of the natural environment including forests, lakes, rivers and wild-life in the first ten classes. The Central Government shall get text books written for the said purpose and distribute them to the educational institutions free of cost, Children should be taught about the need for maintaining cleanliness commencing with the cleanliness or the house both inside and outside, and of the streets in which they live, clean surroundings lead to healthy body and healthy mind. Training of teachers who teach this subject by the introduction of short-term courses for such training shall also be considered. This should be done throughout India".

Remarkably, Supreme Court has thus, opened the door for realisation of fundamental duties with regard to preservation of environment to both government and citizen. Judicial dynamism with regard to environmental matters are appeared due to failure of executive's duty towards protection of environment. The dynamism of an institution is an indication of its responsiveness and effectiveness. When the executive fails in its constitutional duties and is considered to be suffering from "executive inactivism," there is no other recourse for the people but to turn to the judiciary.

Judicial responses in various cases raised the opinion that law is what the law does. Law is meant to eradicate inequalities and human sufferings. There should be a social goal for implementation of laws. Judiciary in India therefore has interpreted fundamental rights as enshrined in the Indian Constitution as a repository of human rights and the fountain source of law for a just society. The right to life as enshrined in Article 21 of the Constitution has been expanded to right of healthy environment. This interpretation is done by Supreme Court in Delhi Gas leak case. The apex court in this case expressed its concern over other hazardous technologies carrying in Delhi and cumulatively violating the standard of life. It lays as:

> There are many other plants in Delhi which are employing hazardous technology or are engaged in manufacture of hazardous goods and if proper and adequate precautions are not taken they too are, likely to endanger the life and health of the community.

Further, in the same case Supreme Court has declared that right to life is incomplete without right to healthy environment. The Supreme Court stated:

> This is a question which needs serious attention of the government of India and we would request the government of India to take the necessary steps at the earliest, because the problem of danger to health and well-being of the community on account of chemical and other hazardous industries has become a pressing problem in modern industrial society.

Thus, Delhi Gas Leak case, for the first time raised the question of far reaching importance concerning the true scope and ambit of Article 21 vis-a-vis the private corporation. The apex court was requested to evolve norms and principles for determining the liability of large private enterprises engaged in manufacturing and sale of hazardous products and the basis on which damages in case of such liability should be qualified. The Supreme Court has assumed violation of the right to life under Article 21 in this case.

It is worth mentioning that in a developing country like India facing with problems of poverty, illiteracy and socioeconomic inequalities, judiciary cannot be concerned merely with the isolationist character of human rights. But this norm is first time broken and environmental pollution was controlled through the apex court in *Rural Litigation and Entitlement Kendra Dehradun V State of U.P.* It is discernible from the attitude of the apex court in this case that on the one hand it realised the importance of the preservation of environment and ecological balance and on the other hand, it also recognised the rights of the affected persons. In this case order of closure though has caused loss of large investment of money, time and effort and also caused hardship to them.

Supreme Court has observed that:

> "It is a price that has to be paid for protecting and safeguarding the right of the people to live in healthy environment with minimal disturbance of ecological balance and without avoidable hazard to them and to their cattle, homes and agricultural land and undue affectation of air, water and environment."

The judiciary has played a pivotal role in interpreting and expanding the scope of fundamental rights, especially that of 'Right to life' enshrined in Art. 21 of the Constitution to establish human rights.

In *Ratlam Municipality v. Verdichand,* Supreme Court has announced healthy environment as a right to life which must be respected at any cost. The apex court hold that 'even as human rights under part III of the Constitution, have to be respected by the state regardless of budgetary provision'.

It further holds that: "Decency and dignity are non-negotiable facts of human rights and are a first charge on local self governing bodies. Similarly providing drainage systems—most pompous and attractive but in working condition and sufficient to meet the needs of the people cannot be evaded if the municipality is to justify its existence".

Thus the Apex Court has strengthened Art. 21 in two ways. First it required laws affecting personal liberty to also pass the tests of Art. 14 and 19 of the Constitution, thereby ensuring that the procedure depriving a person of his or personal liberty be reasonable, fair and just. Second, the court recognised several unarticulated liberties that were implied by Art. 21. It is by this second method that the Supreme Court interpreted the right to life and personal liberty to include the right to a wholesome environment.

The fundamental right to life of people in Article 21 of the Constitution has preferred oversight to life of individual. This view is supported by justice Singh's concluding observations in Ganga Pollution case, where he justifies the closure of polluting tanneries are as:

> "We are conscious that closure of tanneries may bring unemployment, loss of revenue, but life, health and ecology have greater importance to the people".

Regarding maintainability of the petition in Ganga Pollution Case Supreme Court has allowed by giving the reasons of its nature of protecting life of other... Supreme Court in this case observed:

> "The petitioner in the case before us is no doubt not a riparian owner. He is a person interested in protecting the lives of the people who make use of the water flowing in the river Ganga and his right to maintain the petition cannot be disputed".

Apart from it, the Supreme Court has imposed pollution fine on polluters under 'polluter pays principle' and has provided compensation to the citizens affected.

Obviously, from above few cases we can gather the findings that scope and ambit of right to life also explained even on the ground of private corporation. Directive to stop polluting the river stream is issued because of the fact that the case involved the restitution of right to life (Article 21), implementation of Social Policy (Article 48-A) and enforcement of fundamental duty. It was alleged that leather industry discharges waste water containing putrescible organic and toxic inorganic materials in sewage which will deplete and dissolve the Oxygen content of the receiving water course, resulting in emanation of foul odour and the death of all acquatic life. Lastly, Supreme Court by considering nuisance of river pollution as a public nuisance has been given a kiss of life a question of the application of the provisions of Article 32, 21, 48-A, and 51-A(g) i.e. the correlation of fundamental human rights to human environment.

International and national studies of human environment and law reveals that there is a need to develop human environment rights code. It is seen that there is no dearth at international and national level to pay the foundation of human environmental law. It can be concluded that policy goals enforcement of environment is very much present in the judicial responses.

Therefore, it becomes evidently clear and certain that for the declaration of the right to healthy environment as a human right we need effective, efficient environmental laws. On the basis of some cases Supreme Court has acted as a saviour of environmental rights. Significant contribution to environmental causes is made by higher courts and they show much dynamism in respect of protection of environmental rights.

Dynamism of courts is obvious because they themselves stand as a protector of human rights. Peoples' participation is necessary for effective results in this direction. Citizens of India may contribute to this task in a way to secure other area, which still remain unpolluted, by launching a movement. Various non-governmental organisations may also support to the government as well as judiciary to reach upto real problem of pollution. Human rights organisation like Amnesty International and Peoples Union of Civil Liberties (PUCL) may take up the environmental issues like other issues of civil liberties.

As a remedial step, for rapid and sincere solution of environmental pollution the National Environmental Commission and Environment Ombudsman is needed. Judiciary has issued directives for affirmative action by various agencies and instrumentalities to make environment healthy and clean. Ways and means should be devised for follow up of these actions. Environmental awakening among the people will make the environmental laws more enforceable and also lead to develop more purposeful laws in this subject. Sense of environmental rights and duties will make right to healthy environment more strengthful human right.

Thus, for ensuring a healthy environment, traditional as well as modern scientific strategies must be adopted. As we know industrialisation, improper sewage systems, indiscrete use of automobiles, denuding forests, mining and quarrying, slush and burn (jhuming) cultivation pattern, unwise tourism activities in mountains, rivers and seas and other similar so-called developmental activities are mainly responsible for environmental and ecological problems degradation, we must check these. If we speak of healthy environment, we must remember that we have certain accountabilities also. Think not of yourself only, but think of posterity also.

As in the present book, we intend to highlight environmental human rights, in the subsequent chapters we will address all concerned issues citing instances from different countries.

REFERENCES

(1) Adams, Simon; David Lambert (2006). *Earth Science: An illustrated guide to science*. New York NY 10001: Chelsea House. pp. 20.

(2) "Earth's Energy Budget". Oklahoma Climatological Survey. 1996-2004. http://okfirst.mesonet.org/train/meteorology/EnergyBudget.html. Retrieved 2007-11-17.

(3) Oldroyd, David (2006). *Earth Cycles: A historical perspective*. Westport, Connicticut: Greenwood Press.

(4) Simison, W. Brian (2007-02-05). "The mechanism behind plate tectonics". http://www.ucmp.berkeley.edu/geology/tecmech.html. Retrieved 2007-11-17.

(5) Smith, Gary A.; Aurora Pun (2006). *How Does the Earth Work?*. Upper Saddle River, NJ 07458: Pearson Prentice Hall. pp. 5.

2

Why Sustainable Development?: Issues and Concerns

> "Every man is to be respected as an absolute end in himself: and it is a crime against the dignity that belongs to him as a human being, to use him as a mere means for some external purpose."
>
> —*Immanuel Kant*

Sustainable development is a rather recent concept in action though an old in principle. Before describing recent activism for it, we will prefer to put some light on sustainable development.

What is Sustainable Development?

Basically, 'sustainable development' is a pattern of resource use that aims to meet human needs while preserving the environment so that these needs can be met not only in the present, but also for future generations. The term was used by the Brundtland Commission which coined what has become the most often-quoted definition of sustainable development as development that "meets the needs of the present without compromising the ability of future generations to meet their own needs." Sustainable development ties together concern for the carrying capacity of natural systems with the social challenges facing humanity.

As early as the 1970s "sustainability" was employed to describe an economy "in equilibrium with basic ecological support systems." Ecologists have pointed to The Limits to Growth, and presented the alternative of a "steady state economy" in order to address environmental concerns.

The United Nations 2005 World Summit Outcome Document refers to the "interdependent and mutually reinforcing pillars" of sustainable development as economic development, social development, and environmental protection.

The concept has included notions of weak sustainability, strong sustainability and deep ecology. Sustainable development does not focus solely on environmental issues. In 1987, the United Nations released the Brundtland Report, which defines sustainable development as 'development which meets the needs of the present without compromising the ability of future generations to meet their own needs.'

Indigenous people have argued, through various international forums such as the United Nations Permanent Forum on Indigenous Issues and the Convention on Biological Diversity, that there are *four* pillars of sustainable development, the fourth being cultural. *The Universal Declaration on Cultural Diversity* (UNESCO, 2001) further elaborates the concept by stating that "...cultural diversity is as necessary for humankind as biodiversity is for nature"; it becomes "one of the roots of development understood not simply in terms of economic growth, but also as a means to achieve a more satisfactory intellectual, emotional, moral and spiritual existence". In this vision, cultural diversity is the fourth policy area of sustainable development.

Economic Sustainability: Agenda 21 clearly identified information, integration, and participation as key building blocks to help countries achieve development that recognises these interdependent pillars. It emphasises that in sustainable development everyone is a user and provider of information. It stresses the need to change from old sector-centred ways of doing business to new approaches that involve cross-sectoral co-ordination and the integration of environmental and social concerns into all development processes. Furthermore, Agenda 21 emphasises that broad public participation in decision making is a fundamental prerequisite for achieving sustainable development.

According to Hasna, sustainability is a process which tells of a development of all aspects of human life affecting sustenance. It means resolving the conflict between the various competing goals, and involves the simultaneous pursuit of economic prosperity, environmental quality and social equity famously known as three dimensions (triple bottom line) with is the resultant vector being technology, hence it is a continually evolving process; the 'journey' (the process of achieving sustainability) is of course vitally important, but only as a means of getting to the destination (the desired future state). However,the 'destination' of sustainability is not a fixed place in the normal sense that we understand destination. Instead, it is a set of wishful characteristics of a future system.

Green development is generally differentiated from sustainable development in that Green development prioritizes what its proponents consider to be environmental sustainability over economic and cultural considerations. Proponents of Sustainable Development argue that it provides a context in which to improve overall sustainability where cutting edge Green development is unattainable. For example, a cutting edge treatment plant with extremely high maintenance costs may not be sustainable in regions of the world with fewer financial resources. An environmentally ideal plant that is shut down due to bankruptcy is obviously less sustainable than one that is maintainable by the community, even if it is somewhat less effective from an environmental standpoint.

Still other researchers view environmental and social challenges as opportunities for development action. This is particularly true in the concept of sustainable enterprise that frames these global needs as opportunities for private enterprise to provide innovative and entrepreneurial solutions. This view is now being taught at many business schools including the Center for Sustainable Global Enterprise at Cornell University and the Erb Institute for Global Sustainable Enterprise at the University of Michigan.

The United Nations Division for Sustainable Development lists the following areas as coming within the scope of sustainable development:

Sustainable development is an eclectic concept, as a wide array of views fall under its umbrella. The concept has included notions of weak sustainability, strong sustainability and deep ecology. Different conceptions also reveal a strong tension between ecocentrism and anthropocentrism. Many definitions and images (Visualizing Sustainability) of sustainable development coexist. Broadly defined, the sustainable development mantra enjoins current generations to take a systems approach to growth and development and to manage natural, produced, and social capital for the welfare of their own and future generations.

During the last ten years, different organizations have tried to measure and monitor the proximity to what they consider sustainability by implementing what has been called sustainability metrics and indices.

Sustainable development is said to set limits on the developing world. While current first world countries polluted significantly during their development, the same countries encourage third world countries to reduce pollution, which sometimes impedes growth. Some consider that the implementation of sustainable development would mean a reversion to pre-modern lifestyles.

Others have criticized the overuse of the term:

> "[The] word sustainable has been used in too many situations today, and ecological sustainability is one of those terms that confuse a lot of people. You hear about sustainable development, sustainable growth, sustainable economies, sustainable societies, sustainable agriculture. Everything is sustainable (Temple, 1992)."

Environmental sustainability is the process of making sure current processes of interaction with the environment are pursued with the idea of keeping the environment as pristine as naturally possible based on ideal-seeking behavior. An "unsustainable situation" occurs when natural capital (the sum total of nature's resources) is used up faster than it can be replenished. Sustainability requires that human activity only uses nature's resources at a rate at which they can be replenished naturally. Inherently the concept of sustainable development is intertwined with the concept of carrying capacity. Theoretically, the long-term result of environmental degradation is the inability to sustain human life. Such degradation on a global scale could imply extinction for humanity.

The UN Environment Programme (UNEP) defines sustainable development as programmes that would "improve people's quality of life within the carrying capacity of the

Earth's life support system." This means meeting the needs of the present generation without damaging the Earth's resources in such a way that would prevent future generations from meeting theirs. Sustainable development also emphasizes equitable development; that is, it -regards bridging the gap between rich and poor countries as an important way of ensuring that the generations of today and tomorrow can meet their needs.

Since the United Nations convened the Conference on Human Environment in 1972 in Stockholm, concern has been mounting over the continuous deterioration of the global environment. Should this disruption of the global ecological and economic balances continue, it would jeopardise the Earth's life-sustaining qualities and eventually lead to both ecological and economic catastrophe.

A few dangers to earth and population are as followes.

1. Pollution is damaging the Earth's air and water. Each year, our industries produce millions of tonnes of hazardous waste which they dump in the seas, landfills, or, worse, in poor countries. These damage the atmosphere, can alter the climate and are poisonous to us as well as to animals.
2. Human activity is killing off the variety of life. There are some 30 million species on Earth - this diversity of life around us is called biodiversity. It is from these biological resources that we obtain clothes, medicine, food and housing. The greatest threat to biodiversity comes from the destruction of rainforests. In the Amazon, a single tree can contain 2000 unique species of animals. Rainforests are also essential in maintaining the balance of carbon dioxide in the atmosphere. However, unmanaged

logging and agriculture along with acid rain are destroying the rainforests.

3. Careless use of land is reducing cultivable land. Over-farming and cutting down forests, often in the face of changing climates, lead to soil degradation. As land becomes less and less fertile, those who farm it or use it for its trees lose their source of livelihood.

Member States of the United Nations stress that economic and environmental degradation are interrelated and that environmental protection in developing countries has to be viewed as an integral part of the development process. At the same time, consumption patterns and production in industrialized countries speed up environmental deterioration. To solve these problems there needs to be new levels of cooperation among nations -a global partnership.

That was the goal of the Earth Summit held in Rio de Janeiro. The United Nations Conference on Environment and Development, or the Earth Summit as it came to be known, was arguably one of the most significant landmarks in sustainable development to date.

In Rio, five major agreements were reached by 172 heads of state. Non-governmental organizations (NGOs) were also crucial in helping to frame the agenda. Each agreement deals with specific concerns such as the climate or the forests. One of them, Agenda 21, is an all encompassing plan of action which includes most of the important steps that need to be taken to achieve sustainable development

Agenda 21 asks governments and their citizens to do following things: recognize the link between environment and development; issues use energy more efficiently and develop renewable energy sources like wind and solar power; give farmers environmental education; plant new forests and replant damaged ones eliminate poverty by helping the poor earn a living in a way that doesn't damage the environment; impose fines on people and industries that pollute the waters prepare national plans for waste management; require that industry adopt safer and cleaner production methods change wasteful consumption patterns.

A high level body, the UN Commission on Sustainable Development, was set up to oversee the implementation of all agreements reached in Rio. The UN has also set up offices to help countries reach these goals by helping governments, scientists and local people share information and technology and by carrying out training programmes so people can learn about sustainable development and how to achieve it.

Regional Development Activities

The five regional commissions play a crucial role in raising the level of economic and social development in their spheres of activity and in strengthening relations among countries, United Nations organizations and other intergovernmental and non-governmental organizations. During the past few years, as the work of the Organization became more decentralized, the regional commissions shouldered even greater responsibilities. In keeping with the system-wide process of reform, the commissions continued to examine their activities, adjust priorities, restructure programmes and reorganize their staffs to increase efficiency and cost effectiveness.

Now we will put light on some such regional initiatives for protecting environment at regional level.

1. Economic Commission for Africa (ECA)

The cross-currents of modest improvement in Africa's economic performance and major reforms in the programmes, organizational structure and management practices of the Economic Commission for Africa defined the orientation of work of EGA under its Executive Secretary, K.Y. Amoako. The Commission undertook studies on a number of issues, including poverty reduction through improved food security; financing and operation of transport and communications; development of sub regional energy supply, marketing strategies for the enhancement of mineral utilization; problems and prospects for large-scale irrigation in Africa; the integration of environmental considerations into agricultural planning and policy strategies; enhancement of public sector performance, especially in the areas of fiscal policies; and improving public expenditure for rural development. The Commission continued its support for regional capacity building programmes, in particular the Second Transport and Communications Decade in Africa and the Second Industrial Development Decade for Africa.

In collaboration with the Organization of African Unity (OAU), progress was made in implementing the Abuja Treaty Establishing the African Economic Community. Efforts in this regard focused on the preparation of various proposals to the Treaty.

An international conference on reviving private investment in Africa: partnerships for growth and development, organized by ECA, was held at Accra in July 1996 and attended by about 650 participants drawn from African public and private sectors, as well as business executives from outside the region. The Conference was a collaborative effort among ECA, UNDP, the Global Coalition for Africa, the African Business Round Table, the World Bank group and bilateral donors such as the Governments of Japan and the Republic of Korea, the Swedish International Development Cooperation Agency, the Overseas Development Administration (United Kingdom) and the International Development Research Centre (Canada). A key event at the Conference was a round-table summit meeting of heads of State and Government or senior political leaders from eight African countries and six business leaders from within and outside Africa.

Another important initiative was the launching of the African Capital Markets Forum - a multi-country advocacy group bringing together members of stock exchanges, leasing companies and financial institutions such as brokerage firms and investment funds. In the context of follow-up activities related to the cycle of world conferences, ECA has promoted collaboration with other United Nations agencies and regional intergovernmental organizations.

Pursuant to the Programme of Action adopted at the International Conference on Population and Development held at Cairo in 1994, ECA convened an expert non-governmental organization workshop on the Dakar/Ngor Declaration and participated in the joint ECAOAU/African Development Bank missions of the African Population Commission to assess country experiences of some Member States in implementing the Dakar/Ngor Declaration and the Cairo Programme of Action. ECA also convened a senior policy seminar on the social impact of HIV/AIDS in households and family in Africa, held at Addis Ababa in October 1995. With regard to the Fourth World Conference on Women, ECA organized a meeting of the Bureau of the Fifth Regional Conference on Women, which prepared the African Platform for the Conference.

The main objective of the meeting was to discuss a framework for accelerating the implementation of the global and regional platforms for action in Africa. The framework will be disseminated to all national machineries, relevant ministries, non-governmental organizations, training and re search institutions for their use in the implementation process. ECA also co-organized the first Global Trade Fair and Investment Forum (Accra, June/July 1996), which offered women entrepreneurs opportunities for networking, skill development, information sharing, market expansion and forging of partnerships for expanded trade and investment. The intergovernmental machinery for monitoring the implementation of the Programme of Action of the World Summit for Social Development in Africa is the Conference of African Ministers Responsible for Human Development, in conjunction with its follow-up Ministerial Committee of Fifteen.

At the first meeting of the Committee (Addis Ababa, November 1995), several Member States presented country reports on the state of human development in their countries. The Steering Committee of the United Nations System-wide Special Initiative on Africa, launched in March, is co-chaired by the Executive Secretary of ECA and the Administrator of UNDP. In this capacity, ECA will lead or co-lead several priority areas, including South-South cooperation, strengthening civil society, strengthening the informal sector and harnessing information technology for development. During the reporting period, ECA undertook major reviews of its programme and budgeting systems, human resources management, skills assessment and communications strategy.

The new programme orientation, outlined in a document entitled "Serving Africa better: strategic directions for the

Economic Commission for Africa", focuses on five major themes: facilitating economic and social policy analysis, ensuring food security and sustainable development, strengthening development management, harnessing information for development, and promoting regional cooperation and integration. There are two cross cutting issues, namely, gender and capacity-building. The process of reform at ECA was marked by extensive consultations, first with the staff of the Commission and then with 40 high-level African experts drawn from government, the private sector, academia and civil society as well as the Bureau of the Commission.

The third stage was a meeting in early April with the Commission's principal partners: representatives of United Nations agencies, bilateral donors from developed and developing countries, and some international foundations. The new directions document was endorsed at the thirty-first session of the Commission and the 22nd meeting of the Conference of African Ministers Responsible for Economic and Social Development (April/May 1996). At its thirty-first session the Commission also adopted the African Information Society Initiative as a framework for accelerating Africa's entry into the global information system. The Initiative was developed by a high-level working group on information and communication technologies for Africa established in 1995 at the request of the Conference of Ministers. During the session, two special events were organized. The first was a panel discussion on the Framework Agenda for Building and Utilizing Critical Capacities in Africa, which completed a two-year cycle of consultations; the final document is to be published later this year.

The second event, held during the ministerial session, was a high-level forum on public expenditure and the poor that examined, in the context of limited Government resources, various principles and measures for targeting public expenditure for poverty eradication and prevention. A number of the Commission's mandated activities were terminated, deferred or curtailed, while several measures were put in place to achieve efficiencies, including rationalization of the use of resources for travel and of meetings and documentation.

2. Economic Commission for Europe (ECE)

Under the direction of the Executive Secretary, Yves Berthelot, the Economic Commission for Europe continues to promote and facilitate better understanding and agreement on policies, norms and practices among countries of the region and to strengthen their integration and cooperation. This is done through policy analysis and dialogue on macroeconomic and sectoral issues; the continued elaboration of conventions, norms and standards; and a programme of assistance to the transition process of the former centrally planned economies.

In the light of new realities in the ECE region, the Commission established at a special session in September 1995, an ad hoc working group to define strategic directions for the Commission's future activities. The Working Group was created pursuant to the decision taken by the Commission in April 1995 to adopt, at its jubilee session in April 1997, a declaration on strengthening economic cooperation in Europe. To obtain the data necessary for the Working Group to complete its task, a questionnaire was sent to all ECE member Governments.

Analysis of replies by the Working Group is under way and negotiations among members are being held. Changes in the programme of work and the organization of the secretariat will be implemented as a result of this reform exercise. ECE intensified its relationships with relevant organizations in the region, such as the Organization for Security and Cooperation in Europe, OECD, the European Union, the European Bank for Recon -struction and Development (EBRD), the Council of Europe and subregional organizations such as the Black Sea Economic Cooperation, the Commonwealth of Independent States, the Central European Initiative and the Council of Baltic States. Relations with the private sector have also expanded, as private enterprises manifest a growing interest in the ECE forum.

For example, during the Commission's annual session in April 1996, a number of chief executive officers from major corporations participated in a round-table discussion with high-level government officials to discuss cooperation for sustainable industrial development. Closer ties to the private sector were also manifested by the business community's increased financial contributions to the ECE technical cooperation trust fund for industry: structural change and development. The economic analysis conducted by ECE is published in a number of annual and occasional publications.

The Economic Bulletin for Europe, vol. 47 (1995) and the Economic Survey of Europe in 1995-1996 continue to provide annual, in-depth analysis of current economic developments in Europe, North America and the States of the former Soviet Union. Both publications give special emphasis to developments in the transition economies of Eastern Europe and the former Soviet Union and to their progress in creating market economies. The Survey was also used by the senior economic advisers to ECE Governments as background for their annual exchange of information on the development of national economies and on policies and programmes., In the area of sustainable development the assessment of the situation in transition countries was continued, in cooperation with the Committee on Environmental Policy, and a workshop on economic instruments for environmental policies: the response of the private sector was organized by the Governments of the Czech Republic, Finland, Hungary, Norway and Romania in cooperation with OECD and FAO. Population projects in the fields of international migration, fertility and reproductive health and population ageing, which were carried out with financial assistance from the United Nations Population Fund during 1992-1995, were completed and assessed by an independent evaluation team. ECE and the Fund also collaborated on the formulation of a population programme for the period from 1996 to 1999, which focused on the follow-up to the International Conference on Population and Development and the European Population Conference in countries with economies in transition. ECE also published issues Nos. 6 and 7 of the semi-annual International Migration Bulletin containing up-to-date information on migration flows, legislation and meetings in the region. During the period under review, ECE further consolidated its work in the Conference of European Statisticians with the publication of its first-ever statistical yearbook, providing unique statistical profiles of the 55 member States, together with comparative data on economic and social life in the region.

In addition, with the help of EUROSTAT, the International Research and Training Institute for the Advancement of Women and Statistics Sweden, a volume of gender-specific statistics was issued. The Commission also established a flow of essential data on the economies of all countries in transition, which are giving high priority to receiving technical assistance from ECE in the statistical area. The trade facilitation programme of ECE developed-further the norms of the United Nations Rules for Electronic Data Interchange for Administration, Commerce and Transport (EDIFACT) and published two new EDIFACT directories. In addition, six ECE trade facilitation recommendations were revised, including a greatly enlarged recommendation on units of measure, winch represents important progress in the updating of existing recommendations.

A substantial site on the Worldwide Web of the Internet was established and provides access to all United Nations/ ECE trade facilitation recommendations as well as extensive information on EDIFACT. A study on trade and investment finance in the transition economies was prepared by the secretariat and used during the session of the Committee on the Development of Trade and by an international forum attended by more than 300 participants representing government, state and private sectors.

To facilitate distribution and use of agricultural standards, a World Wide Web site was established and is being upgraded and completed. Several new documents and studies were produced during the year, and close cooperation continued with relevant organizations, including OECD, FAO, CODEX Alimentarius, the European Union and the Inter- national Organization for Standardization. The secretariat of the ECE Timber Committee completed a study on the projected supply and demand of wood and forest products over the next 30 years, entitled "European timber trends and prospects into the twenty-first century", and began work on the Forest Resource Assessment 2000, a global effort within which the Committee is responsible for temperate and boreal forests. In the field of industry and technology, a review of the major changes in overall science and technology policies in ECE member States was undertaken, and a number of publications were produced, including the annual reviews of the chemical industry and the steel mar ket, World Engineering Industries and Automation: Performance and Prospects, World Industrial Robots, an updated version of the Directory of Chemical Producers and Products, the biannual review of iron and steel scrap, a study on privatization and changing ownership in the steel industry and the Directory of the Steel Industry and the Environment. Work on coal was dominated by issues related to the restructuring of the coal industries, the clean use of coal, and the development of inter national guidelines, classifications and standards for the sustainable ex ploitation and use of solid fuels.

Activities in the area of electric power focused on the economic reform process in Central and Eastern Europe, environmental protection and the interconnection of international power grids. A symposium on coal-fired power generation, the environment and public acceptance was organized. A study on prospects of natural gas and electric power generation and their impact on the development of the gas and electric power industries was prepared. In the area of natural gas, a glossary on national gas resources was completed and a study on measures of gas conservation: household gas consumption and appliances was prepared.

The gas centre, a technical cooperation programme, completed Its first year of operation, In the course of which four training seminars were organized, four publications were issued, and work was initiated on the development of training manuals and a natural gas database. Under the auspices of the Energy Efficiency 2000 project, trade fair business briefing sessions were held in Bulgaria, the Czech

Republic, Slovenia and Switzerland. A preparatory assistance project to develop energy efficiency demonstration zones in Bulgaria, the Czech Republic, Hungary and the Russian Federation, supported by the Global Environment Facility, was carried out.

A joint project with the European Union SAVE programme on energy efficiency standards was initiated. New work on financial mechanisms for energy efficiency investments in the Russian Federation was started in collaboration with EBRD, the World Bank, Russian commercial banks and western countries. An analysis of the energy situation and policies in the ECE region, with particular reference to economies in transition, and an analysis of energy and CO_2 emission policies and projections were prepared. A workshop on the use of solar energy was held at Bet Berl, Israel. A major highlight during the period under consideration was the third Ministerial Conference on Environment for Europe, held at Sofia in October 1995, for which ECE had been entrusted with overall coordination. At the Conference, Ministers confirmed the important role of ECE in this process and mandated it with the coordination of the next Environment for Europe Ministerial Conference, to be held in Denmark in 1998.

During the intervening period, a draft convention will be negotiated under ECE auspices on access to environmental information and public participation in environmental decision-making. With support at the policy level from the Sofia Ministerial Conference, ECE is developing a programme to review national environmental performance and to examine environmental conditions and practices compared to domestic policy and targets and relevant international commitments. The Sofia Conference also called for strengthening the implementation of and compliance with transboundary environmental conventions administered by ECE, in particular on air pollution, water management, environmental impact assessment and industrial accidents. Under the Convention on Long-range Transboundary Air Pollution, administered by ECE, three new protocols are being negotiated on persistent organic pollutants, heavy metals and nitrogen compounds. The air pollution convention has been presented at meetings of the Economic and Social Commission for Asia and the Pacific, UNEP and the World Health Organization (WHO).

With regard to inland transport, ECE contributed to European integration through the elaboration of uniform norms and standards regarding transport, the facilitation of border crossing and the drawing up of coherent infrastructure networks The European road, rail and combined transport infrastructure networks established under the relevant ECE agreements were being extended to Central Asian and Caucasian ECE member States. A European Agreement on Main Inland Waterways of International Importance was adopted which complements the already existing international instruments providing for the development of international infrastructure networks in Europe.

The Commission continued to make substantive progress in the elaboration, updating and revision of technical requirements for the international transport of dangerous goods by road and inland waterway, and their harmonization with those concerning transport of such goods by rail, sea and air, and initiated work on the elaboration of a European agreement concerning the international carriage of dangerous goods by inland waterway. The Commission also made progress in the international harmonization of requirements for the construction of vehicles, covering a wide range of factors relating to active and passive safety, environmental protection, energy consumption and the removal of technical barriers to trade. The agreement under which this work was done was updated to streamline the procedures and increase their efficiency; worldwide application is being sought and a proposal for a global agreement is under consideration. The Commission paid special attention to road safety, concentrating on the international harmonization of road traffic regulations, road signs and signals by updating legal instruments and adopting recommendations.

In the area of transport facilitation, work continues on a major revision of the TIR Convention (1975), undertaken to cope with increasing fraud and organized crime while ensuring that international road transport and trade in Europe will not be hampered by cumbersome and expensive border-crossing procedures. Preparatory work, in close cooperation with the European Union, the European Conference of Ministers of Transport and other international organizations, has been pursued for the regional Conference on Transport and the Environment to be held at Vienna in November 1997.

Sustainable Development stands for meeting the needs of present generations without jeopardizing the ability of futures generations to meet their own needs – in other words, a better quality of life for everyone, now and for generations to come. It offers a vision of progress that integrates immediate and longer-term objectives, local and global action, and regards social, economic and environmental issues as inseparable and interdependent components of human progress.

Sustainable development will not be brought about by policies only: it must be taken up by society at large as a principle guiding the many choices each citizen makes every day, as well as the big political and economic decisions that have. This requires profound changes in thinking, in economic and social structures and in consumption and production patterns.

2009 Review of the EU Sustainable Development Strategy (EU SDS)

In July 2009 the Commission adopted the 2009 Review of EU SDS. It underlines that in recent years the EU has mainstreamed sustainable development into a broad range of its policies. In particular, the EU has taken the lead in the fight against climate change and the promotion of a low-carbon economy. At the same time, unsustainable trends persist in many areas and the efforts need to be intensified. The review takes stock of EU policy measures in the areas covered by the EU SDS and launches a reflection on the future of the EU SDS and its relation to the Lisbon strategy. The review will be complemented by Eurostat's bi-annual monitoring report on sustainable development which will be published later in 2009.

The European Commission adopted in October 2007 the first progress report on the Sustainable Development Strategy (complemented by a detailed staff working paper). According to the report, there have been significant policy developments in some of the seven key priorities identified in the revised SDS of 2006 - including climate and energy - but progress on policy has not yet translated into substantial concrete action. Eurostat has in 2007 published a monitoring report based on an extended set of sustainable development indicators. This report was one of the inputs for the first progress report on the Sustainable Development Strategy.

The European Council in December 2007 welcomed

the Commission Progress Report and insisted on the need to give priority to implementation measures: "Sustainable development is a fundamental objective of the European Union. The European Council welcomes the Commission's first progress report on the renewed EU Sustainable Development Strategy (SDS). It agrees that the objectives and priorities under the seven key challenges contained in that strategy remain fully valid and that the main focus should therefore be on effective implementation at all levels. The renewed EU Strategy and national strategies for sustainable development also need to be linked up more closely. The governance structure and tools of the SDS, in particular in relation to monitoring of progress and best practice sharing, must be fully used and strengthened. The EU's integrated climate and energy policy and an integrated approach to the sustainable management of natural resources, the protection of biodiversity and ecosystem services and sustainable production and consumption are among the drivers for achieving objectives under both the SDS and the Lisbon strategy. The EU must continue to work to move towards more sustainable transport and environmentally-friendly transport modes. The Commission is invited to present a roadmap together with its next Progress Report in June 2009 on the SDS setting out the remaining actions to be implemented with highest priority."

The EU Sustainable Development Strategy (SDS)

The European Council of June 2006 adopted an ambitious and comprehensive renewed SDS for an enlarged EU. It builds on the Gothenburg strategy of 2001 and is the result of an extensive review process that started in 2004. The renewed EU SDS sets out a single, coherent strategy on how the EU will more effectively live up to its long-standing commitment to meet the challenges of sustainable development. It recognises the need to gradually change our current unsustainable consumption and production patterns and move towards a better integrated approach to policy-making. It reaffirms the need for global solidarity and recognises the importance of strengthening our work with partners outside the EU, including those rapidly developing countries which will have a significant impact on global sustainable development.

The overall aim of the EU Sustainable Development Strategy is to identify and develop actions to enable the EU to achieve a continuous long-term improvement of quality of life through the creation of sustainable communities able to manage and use resources efficiently, able to tap the ecological and social innovation potential of the economy and in the end able to ensure prosperity, environmental protection and social cohesion.

In fact, the strategy sets overall objectives and concrete actions for seven key priority challenges for the coming period until 2010, many of which are predominantly environmental:

(1) Climate change and clean energy
(2) Sustainable transport
(3) Sustainable consumption & production
(4) Conservation and management of natural resources
(5) Public Health
(6) Social inclusion, demography and migration
(7) Global poverty and sustainable development challenges.

To improve synergies and reduce trade-offs, a more integrated approach to policy making is proposed, based on better regulation (impact assessments) and on the guiding principles for sustainable development (adopted by the European Council of June 2005). The external dimension of sustainable development (e.g. global resource use, international development concerns) is factored into EU internal policy making and through integration of SD considerations in EU's external policies.

The EU SDS wants to be a strategy for the whole EU. It therefore proposes mechanisms for improving the coordination with other levels of governments and calls upon business, NGOs and citizens to become more involved in working for sustainable development. An example of this is the launch of a process for voluntary peer reviews of national sustainable development strategies, aimed at improving the sharing of good practices.

Education, research and public finance are stressed as important instruments in facilitating the transition to a more sustainable production and consumption patterns. And because monitoring and follow-up are crucial for effective implementation, the renewed strategy contains a strong governance cycle. Every two years (started in 2007) the Commission is to produce a progress report on the implementation of the strategy. This report is to form the basis for discussion at the European Council, which will give guidance to the next steps in implementation.

First Steps

Already in 1997 sustainable development became a fundamental objective of the EU when it was included in the Treaty of Amsterdam as an overarching objective of EU policies.

At the Gothenburg Summit in June 2001, EU leaders launched the first EU sustainable development strategy based on a proposal from the European Commission. This 2001 strategy was composed of two main parts. The first proposed objectives and policy measures to tackle a number of key unsustainable trends while the second part, arguably more ambitious, called for a new approach to policy-making that ensures the EU's economic, social and environmental policies mutually reinforce each other. The central instrument developed for this purpose was the obligation for the Commission to submit each new major policy proposal to an Impact Assessment.

The EU SDS added a third, environmental dimension to the Lisbon Strategy of economic and social renewal. The two strategies are complementary.

The Gothenburg declaration formed the core of the EU's policies towards sustainable development. But these also encompassed other programmes and commitments, such as the commitments made at the 2002 World Summit on Sustainable Development in Johannesburg and the Millennium Development Goals agreed in 2000, as well as global pledges to increase official development aid and to take account of the needs of developing countries in international trade.

This required a sustainable development strategy with a stronger focus, a clearer division of responsibilities, wider ownership and broader support, a stronger integration of the international dimension and more effective implementation and monitoring.

After a broad public consultation from August till October 2004, in February 2005 the European Commission issued a Communication with initial stock-taking and future orientations for the review. Subsequently in June 2005 the European Council adopted a set of guiding principles for sustainable development. In December 2005 the Commission presented a proposal for a reviewed strategy and platform for further action. The Commission's proposal built on the 2001 strategy and advocated a shift in focus to take account of progress made, tackle shortcomings and take account of new challenges. The result was a renewed strategy adopted by Heads of State and Governments at the European Council of 15-16 June 2006.

With wide environmental awareness steps are being taken for protecting and conserving it.

In the next chapter we will describe the role of UNEP in the direction of environmental conservation for survival of all life on the earth.

3

Human Rights and United Nations Environment Programme

It is pertinent to note that the United Nations Environment Programme, headed by Elizabeth Dowdeswell, introduced organizational and structural change in order to improve programme delivery and to respond effectively to the important tasks mandated to the Programme by the United Nations Conference on Environment and Development. UNEP is participating in the United Nations System-wide Special Initiative on Africa and is a member of the Initiative's Steering Committee. As chair of the Working Group on Water, it promotes an equity-led approach to water resources management in African communities under which it is proposed that all future water policies, plans and programmes in Africa should be assessed in terms of their economic viability, environmental sustainability and equitable use.

In Burundi and Rwanda, UNEP and UNDP are pooling resources to support the rehabilitation of the environment damaged by war. At the request of the Government of Rwanda, the two programmes held a sub-regional workshop on sustainable development at Kigali in September 1995. In December 1995, UNEP organized, in collaboration with the Economic Commission for Africa and the Organization of African Unity, the sixth session of the African Ministerial Conference on the Environment.

In the Asia and Pacific region, UNEP co-sponsored with UNDP, the Asian Development Bank and the Government of Thailand, the Ministerial Conference on Environment and Development in Asia and the Pacific, the first event in Asia to bring together environmental ministers with senior industry officials and corporate leaders. The Conference, hold at Bangkok in November 1995, adopted a Regional Action Programme for Environmentally Sound and Sustainable Development, 1996-2000.

In consultation with Governments and media, UNEP also developed the first Environmental Information and Communication Strategy for the Asia-Pacific Region, covering the period 1995-2000. A handbook entitled "Strategies for environmental citizenship" is available to Governments and non-governmental organizations to assist them in their information activities. UNEP participated in the third Ministerial Conference on Environment for Europe, held at Sofia in October 1995, where it was given the responsibility, jointly with the Council of Europe, to guide the implementation of the Pan-European Biological and Landscape Diversity Strategy through a forum and task force.

The results of a meeting on military activities and the environment in Europe were made available to the Sofia Conference. UNEP initiatives in Europe also included the development of an environmental action framework for the integrated management of the Caspian Sea and a review of the implementation of Agenda 21 in the field of water. With the support of UNEP, the ninth Meeting of Ministers of the Environment of Latin America and the Caribbean was held at Havana in September 1995. This meeting was followed up by an inter-sessional committee meeting of Ministers of the Environment (Mexico City, May 1996), at which UNEP was asked to develop a coordination mechanism for all environmeijtal projects undertaken by United Nations agencies at the regional level. UNEP also increased circulation in the region of its bimonthly Tierramerica newspaper supplement, which by the end of 1996 is expected to reach an audience of over 3 million people.

Further, in the Middle East UNEP participated in the peace process through the multilateral working groups on water resources and the environment. It also contributed to inter-agency meetings in support of socio-economic development in the West Bank and Gaza Strip.

In 1995-1996, eight additional States became parties to the Convention on International Trade in Endangered Species of Wild Fauna and Flora, bringing the number of parties to the Convention to 132. Three additional States became parties to the Convention on the Conservation of Migratory Species of Wild Animals, bringing the number of parties to that Convention to 47. In response to the recommendation of Agenda 21 for UNEP to concentrate on coordinating functions arising from the increasing number of international legal agreements and the functioning of the secretariats of environmental conventions, meetings were held in July 1995 and January 1996 that achieved considerable progress on substantive issues of common concern and on common administrative issues. The Governing Council also requested the Executive Director to place UNEP expertise in forest-related issues at the disposal of the Ad Hoc Intergovernmental Panel on Forests of the Commission on Sustainable Development.

UNEP is also the lead agency of the Panel's programme on the needs and requirements of countries with low forest cover, on which a report was submitted to the Panel at its second session (Geneva March 1996).

In the area of biosafety, the UNEP International Technical Guidelines for Safety in Biotechnology were adopted at the Global Consultation of Government-designated Experts (Cairo, December 1995), following a consensus-building process encompassing a wide spectrum of stakeholders, including the biotechnology industry.

A capacity-building programme related to the implementation of the Guidelines has been formulated by UNEP for submission to potential donors. The Intergovernmental Panel on Climate Change, at its 11th plenary meeting (Rome, December 1995), approved the 1995 second assessment report. Written by some 2,000 leading scientists and technical experts from about 130 countries, the assessment report will improve the world's understanding of the causes of global climate change.

UNEP contributed to the Panel's work by integrating climate-related concerns into its activities on biodiversity, desertification control, protection of the marine environment and promotion of sustainable production and consumption. As decided by the Governing Council at its eighteenth session, UNEP, as a partner in the World Climate Programme, coordinates international activities under thrust 3 of the Climate Agenda: Under the general umbrella of the World Climate Impact Assessment and Response Strategies Programme, which is a component of the World Climate Programme, UNEP has, with funding from the Global Environment Facility, carried out country studies on sources and sinks of greenhouse gases, greenhouse gas abatement costing and climate impact assessment and adaptation.

Remarkably, in April 1996, UNEP submitted to the Commission on Sustainable Development at its fourth session a survey of information systems related to environmentally sound technologies as a contribution to the Commission's deliberations on the transfer of environmentally sound technology.

The Commission invited UNEP to continue its work to develop an information systems network on environmentally sound technologies so as to increase compatibility and cooperation among information systems and sources. UNEP was also invited to consider developing and maintaining a catalogue of environmentally sound technology-related information systems, and eventually to make such a catalogue publicly available in printed form or on diskette and through global networks, such as the Internet. A new Charter for Sustainable Tourism has been developed and is ready for implementation. The Charter is the result of the World Conference on Sustainable Tourism (Canary Islands, April 1995). Guided by Agenda 21, tourism associations, industries and Governments designed this valuable tool with the collaboration of UNEP.

As requested by the Governing Council at its eighteenth session, UNEP convened with the Food and Agriculture Organization (FAO) the first negotiation session on the development of a prior informed consent convention for information exchange on international trade of certain hazardous chemicals and pesticides (Brussels, March 1996).

In addition, a meeting of Government designated experts was convened at Copenhagen in April 1996 on further measures for the sound management of hazardous chemicals. UNEP continued convening a Group of Experts on Liability and Compensation for Environmental Damage Arising from Military Activities (Geneva, September 1995, and London, May 1996). This process has produced an agreed set of conclusions on major issues which will contribute to the work of the United Nations Compensation Commission and will promote the development of an international regime of liability and compensation for environmental damage. UNEP is developing a new state of the environment report, the Global Environment Outlook (GEO).

GEO will report on environmental priorities as defined by regional environmental policy bodies, analyse causal relationships, identify emerging issues and present, where possible, alternative policy options for action. The GEO report series will be produced through a cooperative, participatory and region-based assessment process fully involving regional policy makers, institutions and experts linked by a global network of approximately 20 collaborating centres. UNEP collaboration with the insurance sector resulted in the signing of the Statement of Environmental Commitment by 23 major insurance firms. An insurance statement of commitment was launched in November 1995 and the first workshop to develop modalities for implementation was held in London in May 1996. The agreement, co-signed and supported by UNEP, expresses the industry's commitment to sustainable development. In follow-up activities related to the Fourth World Conference on Women, UNEP has started work with United Nations and non-governmental organization partners to design guidelines for gender sensitive desertification control projects and to promote women's indigenous, knowledge as it relates to biodiversity. issues.

The UNEP Global Youth Forum (San Francisco, United States of America, August 1995) was the largest ever, bringing together 459 delegates from 60 countries. More than 800 children from 83 countries participated in the International Children's Conference (Eastbourne, United Kingdom, October 1995), where, for the first time, children from around the world came together to communicate their concerns about the world environment. The Governing

Council of UNEP at its eighteenth session approved a programme activity budget of $90 million-$ 105 million for the biennium 1996-1997.

The demands placed on UNEP since the United Nations Conference on Environment and Development in 1992 have not been matched by any significant increase in financial resources. The further expected reduction in voluntary contributions to the Environment Fund of UNEP and the unpredictability of payments constitute principal constraints for the future of the Programme and its ability to provide an effective service to the international community.

Issue of Biosphere Protection

It is pertinent to note that Ecology and environment - protection undertakings that have a presumptive claim to the status of world interests are those that appear to be required in order to maintain the basic biospheric conditions that sustain healthy human life on earth. Yet elementary prudence requires that policies be instituted to reduce the likelihood of least those scientifically plausible dire consequences that the majority among the relevant scientific authorities believe could indeed be irreversible.

Currently, there is sufficient (even though far from universal) scientific consensus on the proposition that irreversible dire disturbances to the Earth's life-sustaining ecologies are likely to result from a continuation of certain agricultural and industrial processes, and that the abatement of" these processes now could substantially reduce the risks.

The identified crucial and potentially irreversible biospheric disturbances are ozone depletion and global warming. Moreover, scientific opinion on particular kinds of agricultural and industrial processes that need to be stopped or controlled in order to substantially reduce the most dire risks has also jelled to the point where the domestic and international policy imperatives have become clear.

Ozone Depletion Prevention

We all know that the trace gas, ozone, is found throughout the earth's atmosphere. At lower levels it can be noxious; but in one of its locations—the stratosphere six to thirty miles in altitude-it performs the crucial life-sustaining function of absorbing certain wavelengths of the sun's ultraviolet radiation that can cause cancer and other profound damage and mutations in plant and animal cells.

It is now known that various gases produced on earth by humans have been migrating up to this stratospheric region where they are dangerously thinning the protective ozone layer.

The principal culprits are chlorofluorocarbons (CFCs), one of modern industry's wonder chemical compounds, that have been widely used as coolants for refrigerators and air conditioners, in insulating and packaging materials, and as spray propellants. Understandably, scientific findings about the dangerous properties of CFCs as well as proposals to prevent their production were initially opposed by powerful industry groups and governments representing them. Yet a majority of countries have signed on to the Montreal Protocol of September 1987 and follow-on international instruments committing themselves to stringent CFC abatement policies. The way the international obligations to protect the ozone layer materialized can be instructive to other efforts to achieve cooperation among nation-states in the service of world environmental and ecological interests.

The alarming results of a succession of investigations by respected international panels of scientists in the 1970s and 1980s generated popular pressures that resulted in CFC-abatement legislation in the United States and a few other industrial countries-this despite the fact that no government wants to impose extreme CFC prohibition measures on its own industries without a reliable commitment by other countries to similar measures, considering the costs of developing substitutes for the CFC producing items could reduce the ability of home industries to compete in key sectors of international commerce. But having to bear the costs of the new controls, and determined to rectify their competitive disadvantages, some 500 of the affected U.S. companies formed the Alliance for Responsible CFC Policy to lobby for international controls on CFCs.

Following suit, the United States government changed from an opponent of international abatement treaty to a supporter. In 1986 and 1987 the U.S. Department of State disseminated information worldwide through all of its embassies on the threat CFCs posed to the ozone layer. A similar dynamic operated in other countries where strong environmental movements were bringing on anti-CFC legislation. The affected industries became champions of international controls on CFCs and induced their governments to apply major diplomatic pressures on those countries without domestic controls. Meanwhile the United Nations Environmental Program conducted an intensive program of international scientific and policy workshops on the issue, at which the groundwork was laid for negotiating the formal control protocol.

Trauma of Greenhouse Effect

Although at least as compelling as the need to prevent ozone depletion, it is likely to be more difficult to obtain the international cooperation required to prevent a dangerous warming of the planet from the buildup of concentrations of carbon dioxide (CO_2) and other gases in the atmosphere. This is because the correctives prescribed by most meteorologists would involve drastic, and initially very expensive, changes in the energy production and consumption patterns of most countries-in particular, a conversion from energy obtained from the burning of fossil fuels (principally petroleum and coal) to clean sources such as solar energy. Such a conversion is understandably strongly opposed by many powerful domestic and transnational interests and, in any event, will not occur without government subsidies: Nonetheless, and despite resistance by the Reagan and Bush administrations, a political dynamic analogous to that which produced the ozone protection accords has been gaining momentum.

The governments of various Western European countries and Japan, responsive to domestic environmental groups and industries that anticipate marketing advantages for new energy-efficient and clean products, are beginning to subsidize the needed conversion. As a result, the United States, the largest source of the CO2 buildup, is being subjected to increasing international as well as domestic pressures to stop dragging its feet. Other huge energy users-Russia (highly dependent on petroleum and coal) and China (in the process of converting from coal to petroleum) will have to be brought along if the greenhouse effect is to be reversed. But these and other countries, lagging behind the West and Japan in industrialization, are resistant to converting to clean energy processes without technological and financial assistance from the affluent industrial countries.

In addition, Brazil and other Third World countries engaged in major deforestation projects (the fewer trees the less natural CO_2 absorption) will have to be induced to change their national development plans. One approach, now at the center of international negotiations and on the agenda for the 1992 Earth Summit (see Chapter 6), allows that if a poor country's adherence to new environmental obligations would constrict its economic growth or foreign exchange earning opportunities, such a country should qualify for special technological and financial assistance.

This can take the form of direct help in converting to new technologies or indirect compensation in the form of debt forgiveness or general budgetary support for new "sustainable development" plans and projects. Increasingly, international negotiations on crucial global biospheric protection matters, such as ozone depletion and global warming, are premised on the need and legitimacy of providing such compensation to poor countries who are asked to make costly alterations in their industrialization programs. Although the reasons for sharing international responsibility for the condition of the biosphere are becoming more widely accepted, the translation of this growing consensus into effective policies compels the politically powerful actors in world society to give unprecedented attention to issues of global distributive justice and mutual responsibility.

In the next chapter we will address the burning issue of environmental justice in the context of human rights.

4

Environmental Justice and Human Rights

> "One man can completely change the character of a country, and the industry of its people, by dropping a single seed in fertile soil."
>
> —*John C. Gifford*

To begin with it is apt to quote K.P. Saksena, who says: "Many of the actions of the present generation may impose serious environmental burdens on future generations. By exhausting resources each generation narrows the range of resources available to its descendents. The way we dispose of waste affects the quality of the environment that we pass on to our children and their children. If we reap short-term benefits from cheap disposal of wastes in the air, land, fresh water, seas or even outer space, we may pass on staggering costs on to future generations which must either try to cleanse the disposal sites of abandon them altogether. We also degrade environmental quality for future generations by destroying forests, soils and watersheds which provide essential environmental services. Forest, for example can prevent soil erosion, facilitate the recharge of the water table, harbour wildlife, maintain soil productivity and buffer climate."

The United Nations Conference on the Human

Environment, held at Stockholm in June 1972, proclaimed the right of human beings to a healthy environment and their responsibility to protect and improve that environment for future generations. It also adopted an Action Plan to be undertaken by Governments and international organizations. The opening day of the Conference, 5 June, is celebrated annually as World Environment Day.

Later in 1972, the General Assembly created the United Nations Environment Programme (UNEP) to monitor the environment and to encourage and coordinate sound environmental practices. The first United Nations agency to be based in a developing country, UNEP has its headquarters in Nairobi, Kenya. UNEP's main job is to catalyse environmental action and awareness world-wide. It co-ordinates the environmental activities of all United Nations agencies and works with Governments, the scientific and business communities, and non-governmental organizations, including women's and youth groups.

UNEP has some 200 professional staff and an annual budget of about $40 million. Its activities so far have cost about $310 million and have induced Governments and others to contribute an additional $1,275 million for the protection and improvement of the environment. Growing international concern for the environment is expanding UNEPs tasks and opportunities.

Among UNEP's comprehensive programmes are the Global Environment Monitoring System (GEMS), the International Register of Potentially Toxic Chemicals (IRPTC), and a worldwide data network, Infoterra. UNEP programmes also include a wide range of environmental management activities and supporting services in environmental law, education and information. GEMS provides information in its key areas of climate and the atmosphere, the oceans, renewable terrestrial resources, trans-boundary pollution, and the health consequences of pollution. There are GEMS activities in 142 countries. There are now more than 80,000 chemicals in world-wide use, and through IRPTC, UNEP provides vital information for decisions regarding chemical safety. IRPTC draws on a wide variety of sources, is developing its own data-bank, and has helped 10 developing nations to establish their own chemical information systems. UNEP is also co-operating closely with industry in the fields of safety and pollution control.

Further, UNEP's regional seas programmes help bring countries together to protect common seas. More than 120 countries take part in 10 of these programmes. UNEP has been a catalyst for more than 20 international agreements on the protection of the environments of regional seas, and is currently applying this experience to river basin management in Africa. UNEP has had substantial influence on environmental treaties and conventions. The Montreal Protocol, negotiated under UNEP's auspices in 1987, is designed to help reduce chemical damage to the atmospheric ozone layer, which shields life on Earth from harmful ultra-violet radiation. The Basle Convention, negotiated by UNEP in 1989, provides controls on the international movement and disposal of hazardous wastes.

As of now UNEP is also working with the World Meteorological Organization and Governments on international measures to limit global climate change. UNEP, with other United Nations agencies and non-governmental organizations, is also working on implementation of an Action Plan for the conservation and restoration of tropical forests whose protection is essential in limiting climate change and maintaining species diversity.

UNEP is also active in the areas of desertification control, genetic resources, water conservation, and environmental education. It promotes environmentally sound plans and actions by developing countries to sustain the productivity of their natural resources base, including forests, species, soils and water.

Many nations have improved environmental protection as a result of UNEP's catalytic work, and more than 30 have established national conservation strategies aimed at sustainable economic development.

Earth Summit +5

Action to promote sustainable development worldwide was the focus of a special session of the United Nations General Assembly when it met in New York from 23 to 27 June 1997.

Earth Summit +5, as the special General Assembly session is called, is expected to draw the attendance of heads of State and Government. It will review and appraise improvementation of Agenda 21, and other commitments adopted by the UN Conference on Environment and Development, popularly known as the Earth Summit, held in Rio de Janeiro, Brazil, in 1992. Agenda 21 contains strategies for preventing environmental degradation and for establishing a basis for a sustainable way of life on the planet into the twenty-first century.

"The special session should take a hard, honest and critical look at what has been done and what has not been done since Rio", says Mr. Razali Ismail, President of the General Assembly. "We need to recall and re-emphasize the compact that brought about the Earth Summit'.

The Earth Summit review will assess global progress

made in sustainable development since Rio: Show that sustainable development works by highlighting stories of successful efforts being made by people around the world Identify reasons why goals set in Rio have not always been met and suggest corrective action; Highlight special issues such as finance and technology transfer, patterns of production and consumption, use of energy and transportation, scarcity of freshwater-and identify priorities for future action; Call on Governments, international organizations and major groups to renew their commitment to sustainable development.

In 1992, the representatives of the 172 Governments attending the Earth Summit agreed that urgent action was needed to promote economic and social development while preserving and protecting the environment.

Three Key Documents Were Adopted at the Summit

Agenda 21, a comprehensive plan for global action in all areas of sustainable development

The Rio Declaration on Environment and Development, a series of principles defining the rights and responsibilities of States.

The Forest Principles, guidelines for the sustainable management of forests worldwide.

The United Nations established the UN Commission on Sustainable Development (CSD), which monitors and guides implementation of Agenda 21 and other Rio commitments.

As a Result of the Summit, the United Nations

Convened a Global Conference on the Sustainable Development of Small Island Developing States (May 1994)

Negotiated the UN Convention on Desertification (which

enters into force on 26 December 1996). Negotiated the UN Agreement on Straddling and Highly Migratory Fish Stocks (opened for signing on 4 December 1995). In 1995, the UN Intergovernmental Panel on Forests was set up by the Commission to promote the sustainable management of forests worldwide.

Two international legal agreements the UN Convention on Biodiversity and the UN Convention on Climate Change were opened for signing in Rio and have since entered into force.

The UN Commission on Sustainable Development is at the centre of preparations for Earth Summit +5 and related events.

Two official preparatory meetings were held in New York in 1997:

> The Ad Hoc Intersessional Working Group of the Commission on Sustainable Development, from 24 February to 7 March. The Fifth Session of the Commission on Sustainable Development, from 7 to 25 April. At these meetings, Governments will negotiate documents to be adopted by the special General Assembly session.

Currently, number of events are being organized by Governments, UN agencies and major groups around the world to assess progress in preparation for the special General Assembly session.

Remarkably, the first 1992 Earth Summit in Rio de Janeiro was enriched by the participation of non-governmental organizations (NGOs) and major groups in numbers never before seen at United Nations events. Their involvement reflected the importance attached to the role of civil society in sustainable development, a role that continues to be emphasized by the Commission on Sustainable Development.

Agenda 21 identifies nine "major groups" as partners with Governments in the global implementation of the Rio agreements. They are:

(a) women

(b) farmers

(c) young people

(d) trade unions

(e) business and industry

(f) local authorities

(g) scientists; and

(h) indigenous peoples.

NGOs working in environment and development

During the official 1997 preparatory meetings held in New York, a number of parallel events were organized by these major groups. The April meeting of the Commission included a week-long series of dialogue sessions with major groups. Special panel discussions, exhibitions and other activities were planned during the special General Assembly session in June the same year.

Success stories examples of people around the world taking steps to implement Agenda 21 and make their communities more sustainable. An analysis by the UN highlighting major trends in sustainable development over the past 35 years and their implications for the future.

Comprehensive report on assessment of achievements and failures since the Rio Conference, identifying emerging priorities for the next century.

The roots of Earth Summit +5 reach back to the UN Conference on the Human Environment in 1972, which put environment on the international agenda for the first time. By 1983, the relationship between economic development and its impact on the environment had become the subject of inquiry by the UN World Commission on Environment and Development, known as the Brundtland Commission.

In its 1987 report, Our Common Future, the Commission defined sustainable development as "that which meets the needs of the present without compromising the ability of future generations to meet their needs'" and called for strategies for integrating environment and development As a result, the UN General Assembly decided in 1989 to hold a conference that produced these strategies, and the UN Conference on Environment and Development, or Earth Summit, was held in Rio de Janeiro from 3 to 14 June 1992.

Environmental Justice Movement

It is apt to mention that numerous studies dating back to the early 1970s document that lower-income persons, working-class individuals, and people of color are exposed to elevated health risks in their homes, in their communities, and on their jobs. People of color have borne greater health and environmental risk burdens than the society at large. This is the case even when income is held constant.

For example, race has been found to be independent of class in the distribution of air pollution, contaminated fish consumption, location of municipal landfills and incinerators, abandoned toxic waste dumps, cleanup of Superfund sites, and lead poisoning in children. Governments have been slow to address environmental justice and equity questions of who gets what, why, and how much, who can afford help and who cannot, why some communities are poisoned and others escape, and why some populations are studied while others get left off the research agenda.

Today, disenfranchised communities are calling for greater input into shaping environmental priorities that were in the past left to a small homogeneous group of scientists, health professionals, government bureaucrats, and environmentalists.

The impetus for getting environmental justice concerns on the national agenda has not come from regulatory agencies or the national environmental groups. The impetus has come from the environmental justice movement. This movement is comprised of a loose alliance of grass-roots and national environmental activists, civil rights and social justice leaders, and academicians who question the foundation of the current environmental protection paradigm.

Remarkably, several important events in the early 1990s brought environmental justice concerns into the national public policy debate.

The following are worthmentioning:

1. The dialogue initiated in 1990 among social scientists, social justice leaders, national environmental groups, the federal EPA, and Agency for Toxic Substances and Disease Registry (ATSDR) around disparate impact and environmental equity;
2. The January 1990 letters from Gulf Coast Tenants Organization, Southwest Organizing Project, Commission for Racial Justice, Southern Organizing Committee for Economic and Social Justice, and several dozen other grass-roots environmental justice

groups to the "Big Ten" environmental groups challenging them to end their racism and elitism, whether intended or unintended;

3. Letters from the "Michigan Coalition" in 1991 prompting the federal EPA to form a Work Group on Environmental Equity. The agency later created an Office of Environmental Equity, an Environmental Equity Cluster (coordinated by an assistant administrator for enforcement), issued an Environmental Equity report in 1992, and agreed to meet with environmental justice leaders on a quarterly basis;
4. ATSDR's establishment of a minority health initiative (after some prodding from environmental, health and social justice advocates). ATSDR also held a Minority Environmental Health Conference in 1991 and initiated a health study of communities of color found near National Priority List (NPL) hazardous waste sites;
5. The 1991 First National People of Color Environmental Leadership Summit held in Washington, D.C. The Summit was attended by over 650 delegates from all fifty states (including Alaska and Hawaii), Puerto Rico, Mexico, Canada and as far away as the Marshall Islands. This Summit galvanized grass-roots and national support for strategies to combat environmental racism;
6. The 1992 EPA/ATSDR/NIEHS (National Institute for Environmental Health Sciences) jointly-sponsored "Equity in Environmental Health: Research Issues and Needs" Workshop in Research Triangle, North Carolina;

7. The "Environmental Justice Act of 1992" introduced into Congress by Congressman John Lewis (D-Georgia) and Senator Albert Gore (D-Tennessee). The Act was redrafted and reintroduced in 1993 by Congressman John Lewis (D-Georgia) and Senator Max Baucus (D-Montana); and Appointment of the long-time environmental justice leader Dr. Benjamin F. Chavis, Jr. to head the NAACP, the nation's oldest civil rights organization.

In the US, environmental justice concerns have even permeated the White House. In June 1993, for example, the EPA began drafting a presidential "Executive Order on Environmental Justice" for implementing Title VI of the 1964 Civil Rights Act—a provision that outlaws discrimination in the use of federal funds. The draft Executive Order on Environmental justice calls for an Interagency Task Force to set guidance for social and economic impact reports under the National Environmental Policy Act (NEPA) and Clean Air Act (CAA), equal protection and enforcement under Title VI of the Civil Rights Act, and data collection and analysis on disparate risks and health effects. Environmental justice concerns also are being voiced and actions initiated at the state level. States have begun to recognize the problem and enact legislation to address unequal protection. Arkansas and Louisiana were the first two states to pass environmental justice/equity laws. Virginia passed a legislative resolution on environmental justice. Several other states—including California, Georgia, New York, North Carolina, and South Carolina have pending legislation to address environmental injustices. Texas set up a statewide task force to study environmental racism and disparate impact of its regulations on low-income and people of color communities.

Environmental justice leaders are now challenging those policies and practices that force individuals, workers, and community residents to accept risks others can avoid by "voting with their feet" In 1992 the Southern Organizing Committee for Economic and Social justice held a labor/ environment conference in New Orleans that was attended by over 2,500 grass-roots activists.

Participants at this conference demanded that the problems of pesticides and farm workers, workers in sweat shops (mostly women and immigrants in the garment industry, computer assembly operations, poultry and catfish processing plants, and other dangerous occupations), and lead poisoning of inner-city children be elevated to the national environmental action agenda.

Remarkably, there are other signs that a "new" environmental movement is alive and well in the United States. Grass-roots groups are making their voices heard in Congress, the state houses, and city hall. They are speaking and doing for themselves. Some grass-roots groups (i.e., South west Organizing Project, Gulf Coast Tenants Organization, Mothers of East Los Angeles, West Harlem Environmental Action, Concerned Citizens of South Central Los Angeles, People for Community Recovery, etc.) have begun to form working relationships with national environmental groups (i.e., Citizens Clearinghouse for Hazardous Waste, Greenpeace, Natural Resources Defense Council, Sierra Club, Environmental Defense Fund, National Wildlife Federation, National Audubon Society, etc.), and civil rights and civil libertarian organizations (i.e., National Association for the Advancement of Colored People (NAACP), NAACP Legal Defense and Education Fund, Lawyers' Committee for Civil Rights Under Law, American Civil

Liberties Union (ACLU), Center for Constitutional Rights, etc.).

In 1991 a coalition of environmental, social justice, and civil libertarian groups joined forces to address a childhood lead-screening problem. The Natural Resources Defense Council (NRDC), NAACP Legal Defense and Education Fund, ACLU, and Legal Aid of Alameda County filed a lawsuit against the State of California for not living up to the federally mandated testing for lead of some 557,000 poor children who receive Medicaid. The coalition won an out of court settlement in the *Matthews v. Coye* lawsuit worth an estimated $15-20 million for a blood-lead testing programme. A groundswell of grass-roots organizing is taking place on a host of issues.

Environmental justice has become a major organizing theme for mobilizing grass-roots support. More than 200 environmental justice groups were profiled in the People of Color Environmental Groups Directory (1992). The groups came from thirty-five states, the District of Columbia, Puerto Rico, and Quebec. These and other grass-roots environmental justice groups represent the fastest growing segment of the environmental movement. There can be no doubt that environmental activism is not the sole domain of white middle-class communities.

Environmental justice activists have not limited their attention just to toxics. The same principles of environmental justice as adopted at the First National People of Color Environmental Leadership Summit that challenge ecological destruction and health threats in Love Canal, Three Mile Island, and Times Beach also apply in Triana (Alabama). West Dallas and Texarkana (Texas), in Appalachia, on Native American reservations, and in our urban ghettos and barrios.

Environmental justice does not stop at the U.S. borders. Because of the ethical, economic, and environmental dilemmas posed by the North American Free Trade Agreement (NAFTA), many grass-roots environmental justice leaders actively opposed this agreement. Grass roots groups are demanding an end to unequal protection and environmental injustices. Many of the grass-roots environmental justice groups may not have environment in their name, but they have redefined environmentalists.

Whether they are the Cree in James Bay, the Sioux on the Rosebud, Latinos in the Central Valley of California, African Americans in West Harlem and the Southside of Chicago, or Asian American garment workers in Los Angeles—all are calling for a new environmental protection paradigm—one that has prevention the overarching goal.

The current environmental protection paradigm manages, regulates, and distributes risks. In combination with systemic racism, and a general class bias, the current system:

1. institutionalizes unequal enforcement;
2. trades human health for profit;
3. places the burden of proof on the "victims" and not the polluting industry;
4. legitimates human exposure to harmful chemicals, pesticides, and hazardous substances;
5. promotes "risky" technologies;
6. exploits the vulnerability of economically and politically disenfranchised communities;
7. subsidizes ecological destruction;
8. creates an industry around risk assessment;

9. delays cleanup actions; and
10. fails to develop pollution prevention as the dominant strategy.

Environmental Racism and-Radioactive Colonialism

Many of the American environmental, energy, and military policies distribute costs in a regressive pattern while providing disproportionate benefits for whites and individuals who fall at the upper end of the education and income scale. People of color and the poor have a vested interest in seeing that this nation's environmental and energy policies are just and fair.

In fact, environmental racism is real. It is just as real as the racism found in the housing industry, educational institutions, employment arena, and judicial system. The term of "environmental racism" refers to any policy, practice, or directive that differentially affects or disadvantages (whether intentionally or unintentionally) individuals, groups, or communities based on race or color.

Environmental racism is reinforced by government, legal, economic, political, and military institutions. Environmental racism combines with public policies and industry practices to provide benefits for whites while shifting costs to people of color. Few people would dispute the fact that gasoline taxes are regressive and fall heaviest on the poor. People of color are disproportionately over-represented among the poor. The need for a just and fair energy policy was recently played out in Desert Storm. This "war for oil" fell heaviest on the poor, working class, and people of color who make up a disproportionately large share of America's all-volunteer armed forces. The costs and benefits of defending regressive energy and destructive policies are not equally distributed throughout society.

In the similar vein, the practice of "radioactive colonialism" raises some special ethical and justice questions. Native American reservations are quasi-sovereign nations. Native lands are not subject to federal or state environmental regulations. As a result, more than three dozen reservations have been targeted for landfills, incinerators, and other waste disposal facilities. The vast majority of these proposals have been defeated grass-roots groups on the reservations.

Of the twenty-one applicants for the U.S. Department of Energy (DOE) Monitored Retrievable Storage (MRS) grants, sixteen were Indian tribes. Sixteen tribes lined up for the $100,000 grants from DOE to study the prospect of "temporarily" storing nuclear waste for a half century under its MRS program.

Delegates at the Third Annual Indigenous Environmental Council Networking Gathering adopted a resolution of "No nuclear waste on Indian Lands." Northwest tribes living along the Columbia River-downstream from DOE's Hanford Nuclear Reservation-have experienced the negative side-effects of nuclear waste that have caused mutations in the salmon they depend on spiritually, culturally, and economically.

Environmental Justice Framework Model

The question of environmental justice is not anchored in a debate about whether or not decision-makers should tinker with risk management. The environmental justice framework rests on an ethical analysis of strategies to eliminate unfair, unjust, and inequitable conditions and decisions.

The framework seeks to prevent environmental threats before they occur. The environmental justice framework attempts to uncover the underlying assumptions that may contribute to and produce unequal protection. This framework brings to the surface the ethical and political questions of "who gets what, why. and how much."

The framework includes the following five general characteristics:

1. The environmental justice framework incorporates the principle of the "right" of all individuals to be protected from environmental degradation.
2. The environmental justice framework adopts a public health model of prevention (elimination of the threat beforé harm occurs) as the preferred strategy.
3. The environmental justice framework shifts the burden of proof to polluters dischargers who do harm, discriminate, or who do not give equal protection to racial and ethnic minorities and other "protected" classes.
4. The environmental justice framework would allow disparate impact and statistical weight, as opposed to "intent," to infer discrimination.
5. The environmental justice framework redresses disproportionate impact through "targeted" action and resources.

To sum up, it may be safely surmised that the solution to unequal protection and unjust policies lies in the realm of environmental justice for all. No community, rich or poor, urban, suburban, or rural, black or white should be allowed to become a "sacrifice zone." The lessons from the American civil rights struggles around housing, employment, education, and public accommodations over the past four decades suggest that environmental justice will need to have a legislative foundation. It is not enough to demonstrate the existence of unjust and unfair conditions; the practices that caused the conditions must be made illegal-and ultimately eliminated. How can environmental justice be incorporated into decision making? First, the environmental justice framework demands that the current laws be enforced in a nondiscriminatory way. Second, a legislative initiative is needed. Unequal protection needs to be attacked via federal legislation that moves protection from a "privilege" to a "right." Third, legislative initiatives will also need to come from states. Since many of the decisions and problems lie with state actions, states will need to model their legislative initiatives (or develop stronger initiatives) after the federal legislation.

Fighting for Life

We refuse to accept the deliberate targeting of communities of color and the lands of indigenous peoples as dumping grounds for hazardous wastes and radioactive materials, and the production of pollutants. During the past 20 years, "environmentalism" became a household word. Congress passed landmark legislation designed to protect the air, land and water, many companies embraced pollution prevention and recycling bins became familiar sights.

But not all communities have environmental success stories to share, and not all communities have reaped the benefits of environmental protection equally. The statistics tell a vivid and unsettling story: Three out of every five African and Hispanic Americans live in communities with uncontrolled toxic waste sites. Under the federal Superfund program, abandoned hazardous waste sites in minority areas take 20 percent longer to be placed on the national priority cleanup list than those in white areas.

The poorest African American children are exposed to harmful lead levels at nearly twice the rate of the poorest white children, and the gap widens as family income rises. A more damaging story is told in a report commissioned by National Wildlife Federation's Corporate Conservation Council, Not Just Prosperity, which finds that 98 percent of the major studies environmental inequity show disparities in environmental protection, either race or income.

Not Just Prosperity's findings will provide the background to discussion at the Corporate Conservation Council's annual Synergy conference on February 1. The conference brought together CEOs, government officials and leaders in the conservation and environmental justice movements to address environmental inequity and to recommend actions for improving environmental health in all communities.

No doubt, environmental racism is a fact of American life. All too often, the worst consequences of environmental damage are concentrated most heavily in minority communities. Again too often, resources devoted to correcting the damage are concentrated least where the damage is worst.

Environmental racism exists with its consort, the inequitable influence of environmental degradation upon those who are poor, disadvantaged and least able to fight back. People of color face twice the chance of living close to a hazardous waste facility, as do whites.

They are 60 percent more likely than whites to live in a county listed among the top 2 percent for concentrations of industrial hazards.

In general, the lower your income, the more likely it is you're breathing polluted air.

Non-whites die from acute exposure to hazardous substances at a rate more than 50 percent greater than that for whites. Research findings such as these now compel the debate about our society's response to environmental racism and pollution's disproportionate impact on the poor.

The then Clinton Administration announced a step forward. The Environmental Protection Agency can now use the 1964 Civil Rights Act to investigate complaints of state discrimination in granting permits for the location of such facilities as hazardous waste sites. An executive order was expected requiring all federal agencies to make equal treatment of racial groups a priority in reaching environmental decisions. Congress was considering several bills that would identify "environmental high impact areas" with an eye to declaring a moratorium on new toxic chemical sites there.

The corporate community can act by scrutinizing development plans to assure against locating facilities with high pollution potential in or near communities of color. Redesigned processes that stop the stream of waste by preventing pollution must go forward in plants that already exist.

The pursuit of environmental justice is a paramount priority of the NAACP since it invokes principles of social justice and equal protection. Moreover, environmental hazards pose significant health risks for African Americans and others who are exposed to its toxic effects. Lead poisoning is a classic example of an environmental health problem that disproportionately affects African American children at every income level. Lead affects three to four million children in the United States, most of whom are African Americans and Latinos who live in urban areas.

Among children 5 years old and younger, the percentage of African American children who have excessive levels of lead in their blood far exceeds the percentage of whites in all income levels. People of color communities in America suffer disproportionately from environmental degradation. Specifically, three out of five African Americans and Latinos live in communities with one or more hazardous waste sites.

We believe there is a direct correlation between the disproportionate presence of toxic generating, repository and/or disposal facilities and pollutants in communities where racial minorities and poor people reside and the disproportionate increase in infant mortality, birth defects, cancer, and respiratory illnesses in these communities across the nation. Thus, the issue of environmental racism in our communities has become an issue of life and death.

The convergence of basic survival with the struggle for equality is reflected in the increasing national demand for environmental justice. It is a movement that confronts the immorality of wealthy people consuming the most energy and producing the most waste, while it is the health of the poor that is most affected by the resulting pollution.

It is fueled by findings such as that the Superfund program takes 20 percent longer to place minority communities on the priority list for cleanup, and that the Environmental Protection Agency consistently imposes considerably penalties on those who pollute minority neighborhoods.

Notably, the available research reveals the following facts:

(a) Race proved to be the most significant among variables tested in association with the location of commercial hazardous waste facilities.

(b) Although socio-economic status appeared to play an important role in the location of commercial hazardous waste facilities, race still proved to be more significant.

(c) Communities with the greater number of commercial hazardous waste facilities had the highest composition of racial and ethnic residents.

(d) Three out of every five African Americans and Hispanic Americans lived in communities with uncontrolled toxic waste sites.

(e) There is a clear and compelling need to address the disparate impact of environmental hazards on the health of the African American community, too, must face up to the reality of environmental inequity based on race and income community. This need becomes more graphic when one turns away from the statistics and graphs and looks more directly at the human toll linked to these hazards.

Environmental racism rears its ugly head, for example, in the 80-mile strip along the Mississippi River between New Orleans and Baton Rouge dubbed "Cancer Alley," an area predominantly inhabited by poor descendants of slaves, it is also home to 136 industries, petrochemical facilities and waste dumps that annually release more than 900 million pounds of toxics into the air, ground and water. The inhabitants of that corridor suffer some of the highest cancer rates in America.

A recent example of NAACP involvement in the area of environmental justice is found in the continuing efforts of local residents in Jacksonville, Arkansas to secure protection from the incineration of toxic dioxin-containing wastes stored at the Vertac Chemical Company site.

For several years, the Environmental Committee of the Jacksonville Branch of the NAACP has consulted with environmental experts and lawyers and led the community in the struggle to ensure the health and safety of the predominantly black neighborhoods that adjoin the hazardous site.

The NAACP Legal Department is presently investigating recent requests to join a lawsuit to halt the incineration of toxic wastes and the case looks to be quite compelling.

Tragically, but not surprisingly, those exposed to the highest risks are the non-white residents of lower-economic means in these communities. It is the same pattern that is unfolding all across this country and must be addressed in a serious manner by all who support principles of equal protection.

The environmental and in its internal organization, the environmental community must diversify and practice inclusion. If environmental justice is to be achieved, it can only be on the basis of being indivisible as to race and income.

From the analysis of preceding discussion citing case studies from the USA, it becomes obvious that environmental justice is a grave concern, i.e., fighting for life. In every nook and corner of the world we come to face similar problems, be they developed countries or the developing.

5

Right to Healthy Environment

> "Every indivisual has a place to fill in the world, and is important in some respect, whether he chooses to be so or not."
>
> —*Nathaniel Hawthorne*

Since everybody who comes to earth is bestowed with an environment consisting of air, water, heat, food etc. by the nature, one can demand his/her due share emphatically if denied by others. Hence the right of healthy environment.

In the present chapter we have addressed the issue of right to healthy environment. The description is supported by the case studies of various regions of the world. Sons of soil or the indegenous people as well as the immigrants/ migrants should have this right.

Human rights abuses can not only in the hands of governments by their actions and inactions but also from acts of foreign governments and their instrumantalities. When foreign governments permit their citizens and corporations to export hazardous substances and technologies, without adequate controls and safeguards and without considering the capacity of the receiving countries to cope without unintended and unforeseeable consequences of operation, they are in fact, allowing what one writer describes as 'perverse development' to occur in those countries.

Remarkably, the Chinese, with all their disregard for their own environment, care even less about that of Tibet. Since Tibet is technically part of China, its ecological destruction is ineluctably connected with that of China. Moreover, Chinese are now conducting explorations for oil and other minerals in Tibet, building roads, exporting Chinese settlers, destroying wild-life habitats conducting nuclear tests, dumping nuclear wastes all things which are having a profound impact on Tibet's delicate high altitude ecology. Tons of nuclear waste and other toxic materials must have been discharged by now into the environment by nuclear explosions and reactor spills, in Tibet plateau.

Tibet's population was six millions on its occupation by Chinese, during 1959 and now more than seven million Chinese have been settled there. Overpopulation and industrial technology have contributed in various ways to severe degradation of the natural environment. Such is the deplorable environmental condition of Tibet, whose layman even will care for the smallest worm not to be harmed, because of the impact of Buddhism. Tibetans believe that transmigration of human soul could happen from human to any other sentient being and vice-versa.

The philosophy contains the symbolism of 'mother'. Since one can be reborn in any form through transmigration of soul-which is an ongoing process until the achievement of Nirvana-all other sentient beings have been or have the potential of being ones mother. Hence, the stress on love for them as one's own mother.

Remarkably, the Tibetans lived for centuries in harmony with nature, in their sparsely populated land and professed a philosophy that allowed each first of the nature and animals their rightful place as co-shares. In Tibet's plateau maintenance of the delicate ecological balance is vital for the survival of animal, plant and human life. Today when economics takes precedence even over religion and politics, many governments are under great pressure to increase production substantially and are compelled to take all steps towards high economic growth and to undertake developmental activities on a large scale and thereby the quality of the environment is badly affected. Under such situation there is bound to be more and more strain on the environment and on the depleting mineral and natural resources. It is, therefore, urgent and important that we realise, at earliest, that it will simply be impossible for the sustainance if the strain on it increases. So, for our survival in a clean and healthy environment, it is utmost necessary to reduce our consumption and wastage and, for that, to change our life style.

It is known to some but unknown to many that an average citizen of USA consumes about 20 times more energy and natural resources than an average person in India and China and about 70 times more than a citizen of Bangladesh. So also is the lifestyle of people in Europe. It can thus be imagined that the earth cannot sustain 5.5 billion people at this level of consumption, growth and development. We have, therefore, to adopt a simple life-style if we do not want to perish. After Chinese aggression, as for the Tibetans remaining in their own country are concerned, the odd seem all against them. A few million Tibetans, are faced by the overwhelming and brutal-power and by a vast number, of aggressive, doctrine-infaetuated, rapidly increasing Chinese.

These Tibetans are not permitted to have their share

in the nature and in the name of saving the environment from degradation, all sorts of checks are being imposed upon them. But the spirit of the Tibetans has not been crushed and those who know them will continue to believe that even in the darkest adversity the embers of the faith will be kept alive, as they were before, and something of the Tibetan character with its patience, courage devout simplicity, honest, and kindness will survive to add its rare value to the rich diversity of the world. There are no treaties providing for such a right although thereafter a number of international declarations and proposals which urge governments to recognize and establish the right. Very recently, the rights of future generations has also received some *type* of recognition.

The human right to a healthy and safe environment was first recognized in the United Nations' Universal Declaration of Human Rights of 1948. Article 25 of the Declaration recognizes the right to a standard of living adequate for the health and well-being of an individual. Article 3 can form the basis of a safe and healthy environment since it recognizes the right of life, The, right to a healthy and ecologically balanced environment was reiterated in the 1972 Stockholm Declaration at the United Nations Conference on Environment.

Paragraph I of the Preamble to the declaration states: "Man has the fundamental right to freedom, equality that permits a life of dignity and well-being, and he bears a solemn responsibility to protect and improve the environment of present and future generations" and paragraph I of the Article 25 states: 'Everyone has the right to a standard of living adequate for the health and well-being of himself and his family Neither of these two Declarations are legally binding, but they can provide precedents for the future.

In fact, the right to a healthy and safe environment has also been subject of two recent international documents. The first is the final report of the World Expert Group on Environmental Law, which was established by the Brundtland Commission in 1985. The fundamental right to an environment adequate for the health and well-being of all human beings is contained in Article 1 of the proposed General concerning once rening natural resources and environmental interferences. An Article 2 of the General Principles call upon States to ensure that the environment and natural resources and conserved and used for the benefit of present and future generations.

The second document is the Declaration of the Hague signed by heads of 22 governments in March 1989. "The declaration, which was issued in the context of global climate change states: The right to live is the right from which all the other rights stem. Guaranteeing the right is the paramount duty of those in charge of all States through the world 'Because the problems of warming the global atmosphere and of deterioration of the ozone layer are planet-wide in scope, solutions can only be developed on a global level. The remedies to be sought involve not only the fundamental duty to preserve the ecosystem but also the right to live in a viable global environment and the consequent duty of the nations vis-a-vis present and the future generations to do all that can be done to preserve the quality of the atmosphere'.

The most recent pronouncement on human rights and the environment is contained in Resolution 199040 of 6th March 1990 which was adopted at the 52nd meeting of the Commission on Human Rights of the United Nations Economic and Social Council.

The resolution: "calls upon (governments of all countries of the world) to do their utmost to assist in implementing the right to life through the adoption of appropriate measures at both the national and international levels" and "calls upon all States, appropriate United Nations bodies specialized agencies and inter-government and non-governmental organizations concerned to take the necessary measures to ensure that the results of scientific and technological progress, , are used for the benefit of mankind and for promoting and encouraging universal respect for human rights and fundamental freedoms."

The survey of national and international laws indicated that a human right to safe and healthy environment has not fared well in national and international convenants. State charters of rights and freedom direct too much attention to the rights of individuals vis-a-vis the state when reality society is composed of many groups and the State is not often the most determinative factor.

In fact the rights of an individual against the State are also rights of the individual over other persons and over nature. The very exercise of liberty creates space for domination by some over others and there is no assurance that human rights against the State will not be employed so as cause lawful harm to others. Many of the actions of the present generation may create serious environmental burdens on future generations. By exhausting resources each generation narrows the range of resources available to its descendants.

The way we dispose of waste affects the quality of the environment that we pass on to our children and their children. If we reap short-term benefits from cheap disposal of wastes in the air, land, fresh water, seas or even outer space, we may pass on staggering costs on to future generations which must either try to cleanse the disposal sites or abandon them altogether. We also degrade environmental quality for future generations by destroying forests, soils and watersheds which provide essential environmental services. Forests, for example, can prevent soil erosion, facilitate the recharge of the water table, harbour wildlife, maintain soil productivity and buffer climate. China blasted their largest nuclear test on May 19,1994, at Lop Nor a nuclear testing area which lies directly north of Tibet. China established a nuclear weapons research facility, some three decades ago, on the Tibetan plateau. The Chinese Government has exploded 37 nuclear tests since 1962. The Chinese leaders undoubted look with considerable satisfaction on their achievement in carrying out the nuclear tests despite Soviet Union withdrawal of support in 1960 and active opposition ever since.

During the period 1968-1984, the China is said to have carried out 24 nuclear weapons tests and 150 experimental missile capability Lakes and rivers water provide the environment for many species of organisms and the changes in its temperature by nuclear waste effects the number and kind of such organisms not only in the given body of the water but ultimately affects the human and animal life. China is fast stepping up nuclear energy production to ease power shortages is likely to build its first radio-active waste disposal centre, in the North-West Gansu province, in next couple of years.

The geological surveys and sites for other repositories are being studied. The radio-active waste disposal centre has been necessitated by a mega power plant, China intends to instal at Yamdrok Tso, one of Tibet's lakes which is likely to take Tibet to a gigantic environmental catastrophy. Mr. Martin Frimmel, is assertive and informative when he says: At the holy lake Yamdrok Tso, one of the largest lakes in Tibet, China wants to build a power plant with the Austrian firms-J.M. Voith AG and Elin Energieversorgung (energy supply).

The water is channeled into turbines at Yarluhg Tsampo river four tunnels each 6.4 kilometres long in order to be pumped back into the lake-Yamdrok Tso. The power plant is expected to produce 90 megawatts of energy when operating at top capacity. Mr. Martin Frimmal explains: The ongoing turbulence that would be caused by daily emptying and refilling of the lake water would of course, be of great damage to the sensitive ecology of the lake. The clean, azure blue lake water would be replaced by the dirty water of the river.

The existence of fish population as well as the other animal and plant species is threatened. The process is likely to change the whole ecology and will ultimately effect the animal as also the human life. China being communist, its development activities are secretive in nature. If at any time, the Chemical waste; some toxic waste or radioactive nuclear waste is dumped in the lake, it will lead to an "ecocide" of appealing proportions.

So far as radiation from atomic wastes is concerned, it is not visible. It cannot be touched or felt. But even though hidden from the eyes, the radiations bring about changes in the proteins enzymes, and the genetic factors of the human body. Once affected, the body can never get rid of it. It stops the growth of other live tissues. The power of the blood cells to fight diseases decreases. The blood even gets polluted owing to ground water. The consequences are arthitis, abortion, allergy, heart-diseases, diabetes, cancer, birth of undeveloped Children etc. However, in the atomic wastes, nuclear fission continues but silently and about a period of three centuries their atoms get transformed into highly radio-active plutonium atoms. However, the Yamdrok Tso hydroelectric plant is projected by the Chinese government as a hydro-electric power build in a poor area provide electricity for local people to develop their economy. The project, surrounded by Tibet's snow capped peaks, at the height of 11,000 feet, is said to be aimed at relieving centuries old poverty and backwardness in the land that had no electricity or roads until 1950. The project has significant and it has ethical implications. Tibetans have steadily opposed the project and it is located in what is supposed to be a Tibetan "autonomous" region.

The late Panchen Lama, the second highest monk in Tibetan Buddhism after the Dalai Lama, was an outspoken . of this project because of the harmful cultural environmental implication. Moreover, the Tibetans from their post experience are convinced that almost all the economic benefits of the developments projects are being cornered by the Chinese settlers. Tibet's so called progress is viewed by them for China to the detriment of Tibet. The presence of Chinese settlers since May 23,1951 and more particularly after 1959 has placed new economic and ecological pressures on the resources of Tibet.

Though the Yamdrok Tso Hydro-electric power project involve international funding and technical assistance .'to improve the lot of Tibetan population. But the projects are likely to cater more to the needs of Chinese settlers rather than supporting the needs of the wider Tibetan population in the Tibet Autonomous Region. (TAR). The project if completed will emerge as a living monument of Human Rights violations.

Right to Live in Healthy Environment vis-a-vis Human Excretion

Every one likes to live in the healthy environment which is basic human right. Healthy environment is a nature's gift. Air and water are essential for living things. Human beings cannot survive for a moment without the air, hence the air should be pollution free. But it is very difficult to get pollution free due to faulty drainage system, human excretion, unplanned industrialization, over-population, and vehicular traffic. Environmental pollution is a worldwide problem and India too is facing this menace. The United Nation's Declaration on Human Environment, 1972 cautioned about the environmental pollution as:

We see around the growing evidence of man-made harm in many regions of the earth. Dangerous level of pollution in water, air and living beings, major and undesirable disturbances to ecological balance of the bio-sphere, destruction and gross deficiencies harmful to the physical, mental and social health of man, in the man-made environment particularly in living and working environment. Environmental pollution is natural as well as made by men. Pollution of air and water is man-made and is serious than the, natural pollution, Over-population is the main cause for making the air as well its water Polluted through human excreta. Though we have Air (Prevention and Control of Pollution) Act, 1981 and Water (Prevention and Control of Pollution) Act, 1974, still air and water pollution has not been controlled to the desired extent. Faulty drainage system and dry latrines may be cause for it. Auto-exhaust is another cause for environmental pollution because due to improper maintenance they exhale huge quantity of carbon-monoxide, hydrocarbons, oxides of nitrogen and other pollutants.

Among other pollutants human as well as animal excretion is one of the serious sources for polluting the environment. In most of the cities manual scavengers are engaged to carry the excreta to the nearby open space, river, canal, nullah, etc. In some of the places without constructing septic tank flush pipes are directly put in the canal or nullah which flows through the heart of city. In the absence of public lavatories the slum-dwellers use the nearest open space to ease themselves. Even drains are used as a defecation place by human beings. They live themselves in unhealthy environment as well as make others to live in the same.

United Nations had pointed out that human wastes may cause pollution by seepage from improperly constructed or improperly placed septic tanks, cesspools, leaking sewer lines. Industrial wastes including highly poisonous chemicals, may be introduced to ground water either by intention or accident. Even the dumping and covering of vegetable materials in garbage result in their decomposition and carrying down of decomposition products, including carbon dioxide gas, by percolating water into underground water bodies from the articles.

In Part IV of the Constitution, Article 48-A which was added in 1976 vide 42nd Amendment Act contains the directions for States to protect and improve the environment and safeguard the forests and wild-life. Even fundamental duty is imposed on every citizen of India vide Article 51-A(g) of the Constitution to protect and improve the natural environment including forests, lakes, rivers and wild-life and to have compassion for living creatures. For the survival of the man, he must have a healthy environment Article 21 of the Constitution guarantees a fundamental right to life, a life of dignity to be lived in a proper environment, free of danger of disease and infection. Right to live with dignity is meaningless if one is engaged to carry the human excreta of another over his head or shoulder. It also violates the fundamental right to live in healthy environment of those near whom that human excreta is dumped. Living in the polluted environment is like dying every moment, with eyes, ears, lungs, kidneys, skin, etc. exposed to variety of pollutants.

Also, an enactment namely Employment of Manual Scavengers and Construction of Dry Latrines (Prohibition) Act, 1993 has been made for the purpose of ensuring a healthy environment and to provide right to live with dignity.

Employment of Manual Scavengers and Construction of Dry Latrines (Prohibition) Act, 1993

The object of Employment of Manual Scavengers and Construction of Dry Latrines (Prohibition) Act (hereinafter referred as E.M.S. and C.D.L. Act) is to prohibit the employment of manual scavengers and construction of dry latrines. Manual scavengers and dry latrines are interlinked with each other The purpose of E.M.S. and C.D.L. Act to prohibit the manual scavenging is in consonance with the preamble of the Constitution which provides fraternity assuring the dignity of the individual. It is also in tune with Article 21 of the Constitution. Right to life includes right to live with human dignity as the High Court of Madhya Pradesh held in the case of *Dr. K.C. Malhotra v. State of M.P.* In the instant case the Court also observed that right to live is the fundamental right of every citizen and therefore, in the discharge of its responsibilities to people, State has to provide at least minimum conditions ensuring human dignity. Right to life enshrined in Article

21 cannot be restricted to mere animal existence. It means something much more than just physical survival.

Right to life includes right to live with human dignity, namely, the bare necessities of life such as adequate nutrition, clothing, shelter, facilities of reading, writing, expressing oneself in diverse forms, freely moving about and mixing and commingling with fellow human beings. The right to life and personal liberty includes the right to live with decency and dignity, however person employed to scavenge the human excreta is denied right of decency and dignity. In spite of the existence of this legislation, the dehumanising practice of manual scavenging of human excreta still continues in many parts of India. The other object of E.M.S. and C.D.L. Act is to improve and protect the human environment by making it obligatory to convert dry latrines into water-seal latrines or to construct water-seal latrines in new constructions. The aim of E.M.S. and C.D.L. Act to improve and protect the human environment is in consonance with the directive principles of the state policy regarding the standard of living of its people and the improvement of public health as its primary duties.

The provision to regulate the construction and maintenance of water-to latrines under the E.M.S. and C.D.L. Act is a right approach towards right to live in healthy environment which is a basic right of a person under Article 21 of the Constitution. The E.M.S. and C.D.L. Act is in addition to the Municipal laws. Where the Municipal laws are not stringent enough to prevent the construction of dry latrines or conversion of dry latrines into water-seal latrines this Act will come into play.

Hence, it is necessary to have a uniform legislation for the whole of India for the purpose of eliminating manual scavenging by declaring employment of manual scavengers for removal of human excreta an offence and thereby imposing ban for further proliferation of dry latrines in the country.

The matter relating to health is covered by the State list, hence, State is empowered to enact laws regarding the health. The Indian Constitution is a touchstone relating to any law operating in India which can be tested by it. During two years of its enforcement, the Constitutional validity of the E.M.S. and C.D.L. Act or-any of its provisions has not been challenged in a Court of law. It may be either because of constitutionally perfect law or the span of period between the enactment and its enforcement has been so less that there seems to be little inter-action of the law and the people to give rise to any conflict on the basis of constitution.

It may be because the general awareness of constitutional rights towards the environment is very low. Though the Parliament has no power to make laws relating to health and environment matters, but it seems to have enacted the E.M.S. and C.D.L. Act under the provisions of Articles 249 and 250 of the Constitution. To make the E.M,S. and C.D.L. Act applicable to the States, the legislature of the concerned state has to pass the resolution that particular matter should be regulated in the state by the parliament. Hence, in pursuance of Article 252(1) the Constitution, resolution have been passed by only six States, namely, Andhra Pradesh, Goa, Karnataka, Maharashtra, Tripura and West Bengal that matter relating to prohibition of manual scavengers and construction and maintenance of dry latrines should be regulated by the Parliament by law.

But other States of the country are yet to seek the extension of that Central piece of legislation for the prohibition of manual scavenging and construction of dry latrines. However, this Act is applicable to all the Union Territories of India.

Generally, the Central legislation is applicable to whole of India except the State of Jammu and Kashmir, because it enjoys special status under Article 370 of the Constitution, But E.M.S, and C.D.L. Act firstly is applicable to the whole State of Andhra Pradesh, Goa? Karnataka, Maharashtra, and West Bengal and to all the Union Territories. It shall apply to such other States which adopt E.M.S and C.D.L. Act by resolution passed in that behalf under Article 252(1) of the Constitution. This provision provides the discretionary powers to the States whether to adopt this Act or not which may debar the unification of law because it may take years together to adopt this Act by all the States.

Section 2(c) of E.M.S. and C.D.L. Act defines the dry latrines as latrine other than the water and seal latrine, But it does not cover the vacant places near the buildings or roads used as latrine by inhabitants of jhuggies and direct flush pipes are put in the canal, nullah etc., which also cause the pollution in the environment. Section 2(d) of the Act defines environment which "includes water, air and land and the inter-relationship which exists among and between water, air and land and human beings other living creatures, plants, micro-organism and property.

Section 2(e) says environmental pollutant means any solid, liquidor gaseous substance present in such concentration as may be or tend to be injurious to environment. Further Section 2(f) defines environment pollution as the presence in the environment of any environmental pollutant. The term latrine is also defined by Section 2(i) as a place set apart for defecation together with structure comprising such place, the receptacle therein for collection of human excreta and the fittings and apparatus if any connected therewith.

Now it is a crime to engage or employ a person for manually carrying human excreta from the dry latrines. Construction or maintenance of dry latrine is also punishable. These provisions of E.M.S. and C.D.L. Act recognised the fundamental right to life which includes right to live with decency and dignity under Article 21 of the Constitution.

However, the State Government has to issue a notice of its intention to take action before ninety days. It has also to make sure that sufficient number of water-seal latrines are existing in such area for the use of public. When it is indispensable in the interest of public health for the protection and improvement of environment to do so, under Section 3(1) of E.M.S. and C.D.L. Act. It also recognises the human right to live in the healthy environment which is fundamental right. But the State Government may exempt by general or special order any area, category, buildings or class of persons from the conditions, any provisions of this Act, any rule, order, notification or scheme or dispense with the observance of any such requirement in class or classes of cases, if it is satisfied that compliance with such provisions or requirement is or ought to be exempted dispensed with in the circumstances of the case.

The State Government may appoint a District Magistrate or Sub-divisional Magistrate as an Executive Authority to exercise jurisdiction within such area as may be specified in the order and confer powers and impose duties on him which may be necessary to carry out the provisions of the Act, The executive authority may further specify the officer or officers, subordinate to him, who shall exercise all or any of the powers and perform all or any of the duties, which are conferred or imposed on him in the local limits within which such powers or duties shall be carried out by the officer or officers so specified.

Further, the executive authority shall as far as practicable, try to rehabilitate and, promote the welfare of the portions who were engaged in or employed for manual scavengers in any area for which a notification under Section 3(1) of E.M.S. and C.D.L. Act has been issued by securing and protecting their economic interests.

Section 6 of the E.M.S. and C.D.L. Act empowers the government to make schemes to carry out the provisions of the said Act. The State government may by issuing notification make one or more schemes for regulating conversion of dry latrines into, or construction and maintenance of water seal latrines, rehabilitation of persons who were engaged in or employed as manual scavengers in any area in respect of which a notification under Section 3(1} has been issued in gainful employment and administration of such schemes. Different schemes may be made in respect of different areas and for different purposes of the Act. Proviso attached to Section 6(1) says that such scheme shall be made with the consent of HUDCO where financial assistance is made from it.

In fact, the Scheme may also include the time bound phased programmes for the conversion of dry-latrines into water seal latrines; technical or financial assistance for new or alternate low cost sanitation to local bodies or other agencies, construction and maintenance of community latrines and regulation of their use on pay and use basis; construction and maintenance of shared latrines in slum areas or for the benefit of, socially and economically backward classes of citizens; registration of manual scavengers and their rehabilitation; specification and standards of water-seal latrines, procedure for conversion of dry latrines into water seal latrines; and licensing for collection of fees relating to the community latrines or shared latrines.

Under Section 7 of the E.M.S. and C.D.L. Act the State government is empowered to issue directions in writing for performance of its functions under Act to any person, officer or local or other authority and such person, officer or a local or other authority shall be bound to comply with such directions. The State government vide Section 9(1) may appoint the inspectors for the purpose of carrying out the inspection within the local limits as prescribed under the Act. Among the other powers the inspector under Section 9(3)(c) is also empowered to examine and test any latrines or conduct inspection of any building in which he has reason to believe that an offence under this Act or the rules, orders or schemes made thereunder has been or is being or is about to be committed and to prevent or mitigate environmental pollution.

Vide Section 8 of the Act the executive -authorities, inspectors, officers and other employees authorised to execute a scheme or order made under this Act, when acting or purporting to act in pursuance of any provisions of this Act or the rules or schemes made or orders or directions issued thereunder shall be deemed to be public servants within the meaning of Section 21 of the Indian Penal Code.

Section 10 of the E.M.S. and C.D.L. Act strengthens the hands of the executive authority by conferring the powers on it to prevent environmental pollution. The executive authority on receipt of information with respect to the fact or apprehension of any occurrence of contravention of the provisions of Section 3, whether through intimation by some person or on a report of the inspector or otherwise, the executive authority shall as soon as possible in addition of taking any other action under this Act, direct the owner or occupier of the premises to take such remedial measures, which may be necessary within such reasonable time as may be specified therein. If the owner or occupier concerned fails to comply with such directions, cause such remedial measures to be taken as am necessary to prevent or mitigate the environmental pollution, at the cost of such owner or occupier of the promises. In case any expenses are incurred by the executive authority while carrying out the remedial measures may be recovered from the person concerned along with the interest at the rate specified by the State government with effect from the date when a demand for the expenses is made, such recovery shall be made as arrears of land revenue or of public demand.

"Whoever contravenes the provisions of the Act, rules, orders, direction and schemes shall be punished under section 14 of the B.M.S. and C.D.L. Act. For each such failure or contravention punishment is imprisonment for a term which may extend to one year or with fine, which may extend to two thousand rupees or with both in ease of the failure or contravention continues, with additional fine which may extend to one hundred rupees for every day during which such failure or contravention continues after the conviction for the first such failure or contravention.

If the offence is committed by the company under this Act such company as well as every person incharge of and responsible to the company for the conduct of its business at the time of commission of the offence, shall be deemed to be guilty of the offence and shall be liable to be proceeded against and punished according to law. Proviso to Section 15(1) says that if such person proves that the offence was committed without his knowledge or that he had exercised all due diligence to prevent the commission of such offence. Section 15(2) of the Act lays down where an offence under this Act V has been committed by a company and it is proved that the offence has been committed with the consent or connivance of, or that the commission of the offence is attributable to any' neglect on the part of any director, manager, managing agent or such other officer shall also be deemed to be guilty of that offence and shall be liable to be proceeded against and punished, accordingly. According to explanation attached to this Section company means any body corporate and include a firm or other association of individuals and director in connection to a firm means a partner in the firm.

Offences committed under the E.M.S. and C.D.L. Act shall be treated as cognizable. Section 17(1) of this Act deals with the jurisdiction which says that if any offence is committed under this Act.

Obviously, the preceding description makes it clear that the problem of healthy environment is a world-wide phenomenon. Its grave consequences are being felt everywhere. It is seen that the developed countries are the more accountable for polluting environment thus posing a threat to life on earth.

In the next chapter we will take the issue of environmental pollution in the context of human right.

6

Environmental Pollution and Human Rights

> "Given today's environment, we are living with a time bomb that is going to explode in 20 to 30 years from now in the form of more persons being stricken with cancer."
>
> —*Frant Rauschel*

Our abode Earth Planet appears to be at unrest. Man's ambition for everlasting enjoyment and comfort had led him to exploit natural resources to the extent of reducing its natural capacities for self-stabilization. As a consequence of the outright disregard of the impact of those activities, numerous environmental problems have arisen which seem to be the bane of modem scientific thought and technological advancements. It is only recently that man has come to realise that the Environment is an unbroken totality which includes everything that affects the air we breathe, the water we use for drinking and irrigation purposes, the surroundings in which we live and other factors that affect the quality of life on Earth.

Thus pollution of environment is a world-wide phenomenon. To combat it the United Nations Organisation has also expressed its deep concern. It organised a conference on Human Environment at Stockholm in June, 1972 which laid down the principles and action plans for regulation and controlling the Human Environment and also deliberated on the institutional and financial arrangements for the purpose, specifically the International Conference on Human Environment declared:

(i) man has the fundamental right to freedom, equality and adequate conditions of life, in an environment of quality that permits a life of dignity and well-being, and

(ii) man bears a solemn responsibility to protect and improve the environment for present and future generations.

Further, in June 1992, the United Nations Conference on Environment and Development (UNCED), popularly known as Earth Summit, took place in Rio. More than 170 leaders of countries participated in the Conference and committed their countries to environmentally sustainable development. The Rio Declaration stated:

> ... environmental protection shall constitute an integral part of the development process and cannot be considered apart from it.

Also emerging from the Rio Conference was Agenda 21, a master plan of action for moving towards sustainable developments. The action plan includes sections on poverty, changing consumption patterns, international trade and cooperation, promoting sustainable agriculture, rural development, desertification and deforestation, protection of atmosphere, oceans and the mountains; toxic, hazardous, radioactive and solid wastes, women and development, indigenous people, the role of NGO's and multilateral agencies, etc. etc. The Rio Conference also estimated that about 600 billion a year would be required for effective earth restoration and for its care and maintenance.

Our country, India being one of the signatories to Stockholm Conference took note of all these environmental problems and expressed concern about the environmental issues.

The Government of India for the first time articulated its deep concern in the 4th Five Year Plan (1969-74). The plan emphasised that it was an obligation of each generation to maintain the productive capacity of land, water, air and wildlife in a manner which leaves its successors some choice in the creation of healthy environment. The successive, Fifth, Sixth, Seventh and eighth Five Year Plans also emphasised the need and importance of environment protection.

Notably, the preamble of Indian Constitution declares that social, economic and political justice is to be secured and protected to every citizen of India. Further under the Fundamental Rights, Art. 21 of the Constitution guarantees the right -to life, a life of dignity to be lived in a proper environment, free of danger of disease and infection.

The Directive Principles of State Policy of Indian Constitution in Art. 47 clearly declare to raise the level of nutrition and the standard of living of its people, and improvement of public health as one of the fundamental duties of the State, Article 40(A) which was incorporated by the 42nd Amendment of the Constitution provides that "the State shall endeavours to protect and improve the environment and to safeguard forests and wildlife of the country.

The 42nd Amendment of the Constitution also added

certain fundamental duties of the citizens in Part IV-A of the Constitution. Under these Fundamental Duties, Art. (51A) specifically provides as such............" It shall be the duty of every citizen of India to protect and improve the natural environment including forests, lakes, rivers and wildlife and to have compassion for living creatures. Besides these provisions, there are other entries in the Constitution which deal with the matters concerning environment and its planning and control These provisions of the Constitution empower the Central and State Governments to take legislative and administrative measures to protect, preserve and improve the quality of environment.

It is remarkably that among the factors that upset the balance of nature and jeopardise human life is noise. In order to function efficiently, physically as well as psychologically man needs peace and quiet. But the growing levels of noise in the mechanised society of today have become a hazard to public health. Prolonged exposure to noise may cause hearing loss, interference with speech, communication and sleep, annoyance, reduction in working efficiency fatigue nervousness, irritability, increased blood pressure and many other symptoms.

Excessive noise produced in recent years has compelled the governments to introduce "Insulation Grant Schemes" in U.K. and U.S.A. Medical experts in United Kingdom have estimated that one out of four men and one out of three woman have developed neurosis as a result of exposure to excessive noise, In the same way, every fifth patient admitted to the psychiatric wards of French Hospital lands there because of noise.

According to the Ministry of Health, Canada, two million candidates suffer from poor hearing due to exposure to high levels of noise. The situation is so worse that a leading expert Dr. V.O. Krudson has warned, "If the noise level we live with today continues to increase at the same scale for next 30 years as it has for the last 30 years, it could become lethal". Similar conclusions have been reached by psychologists, psychiatrists, scientists, medical experts, urban planners and other experts in related field.

In fact, high levels of noise adversely affect not only human beings but even animals. Many birds for example, leave the place if It becomes too noisy and go elsewhere to nest. There is a decrease in migratory birds to a place if it is subject to noise. It has been reported by Director of Delhi Zoo animals, particularly the deers, lions and rhinos are worst affected by traffic - noise. They become dull and inactive, their health deteriorates and their number does not increase. According to the Director of Delhi Zoo, the psychological and ecological consequences of noise could be serious from the point of view of their survival. Similarly in a survey in Malaysia, it was observed that grizzly bears, musk-oxen and Kangaroo rats are falling victims to noise pollution.

As per the latest revelations made by Deptt. of Tourism and Development in 1993 Jammu has been placed in the high rankings in traffic concentration per square unit of the road available. The heavy influx of road traffic has made it mandatory to study and analyse the proposition of noise pollution of the same. The city in toto has a population of over 7 lakhs now and also bears the brunt of over 2.8 million Vaishno Devi pilgrims annually. This affects the civil amenities infrastructure of the city. It has an area of over 33 sq. kms. The city has roads/lanes with average 3 meter width as 150 kms; 2 meter with 40 kms; surface drains 250 kms and deep drains/nullah 25 kms. On an average, the city has a discharge of 125 metric tonnes of city waste per day. With a explosion in population in Jammu, the roads have shrunk, resources have been greatly reduced and the vehicular traffic has increased multifold.

The number of registered vehicles by the close of year 1994 was approximately one lakh and fifty thousand which includes buses, mini-buses, trucks, taxis, jeeps, cars, three-wheelers, scooters, tractors, trailers and vehicles of other sort. The chief source of noise pollution in the city of Jammu has been the vehicular traffic. Speed of traffic has been one of the most important causes of noise on the roads. Modern road developments encourage higher speeds which in turn lead to greater noise volume. All motor vehicles produce noise from their gear boxes and exhaust systems. In addition, heavy vehicles also produce rattle squeaks and vibrations according to their age and degree of holding. On any road there is general noise level as long as any traffic is moving. This general level varies with traffic density and time of the day. Distinctive and regular peaks in the noise level occur in the morning and evening as people travel to and from work. Passing of individual vehicles at any given point leads to super-imposition of peak traffic noise levels of a few seconds duration on the general noise level at that point. These peaks vary according to the size, type and speed of the vehicle.

Traffic noise has insidious effects, for example continual exposure to traffic noise may accelerate the age induced deafness. Traffic noise certainly disturbs the quality of sleep and, in a few individuals, it may lead to insomnia. At the very least, traffic noise disrupts relation and enjoyment of life in the home, and reduces the pleasure derived from conversation, listening to the radio and watching the television.

Delhi is said to have largest number of vehicles in the

country: The city is badly affected by noise pollution besides air pollution and water pollution.

In order to overcome the problem of noise pollution caused by automobiles, and to give the basic human right i.e. right to live in a peaceful and healthy environment to the people it is submitted that:

1. Noise can get reduced by prescribing noise limits from vehicular traffic, ban on honking of horns in certain areas and planning wain traffic arteries, residential colonies, fly-over bridges, creation of silent zones near schools and hospitals and redesigning of buildings to make them noise proof. Other measures can involve reduction of traffic density in residential areas giving preferences to mass public transport system like Metro and Busses.
2. Noise pollution caused by automobiles owes much to the uncallous and negligent attitude of traffic person this perspective lack of proper training regarding environment seem to be another cause of increasing noise pollution caused by automobiles. Thus there should be scientific training programmes so that the law enforcement agencies are adequately educated towards controlling checking/eradicating the increasing menace of noise caused by automobiles.
3. There should be plantation of trees of noise absorbing properties like Ashok, Neem, Tamarind, etc. alongside roads as a measure of reducing noise pollution.
4. There is a growing realization that voluntary agencies which are involved in various development work also play an important role for the protection of the environment. There are at present different Voluntary Agencies/Social Action Groups who are

doing a good job for the protection of the environment. But in the field of noise pollution, neither of them had done any appreciable work.

5. The awareness amongst masses is essential. People can be educated through radio, TV, news reels in cinema halls about noise pollution. Public awakening is also very essential. In India, most of the persons seem to be ignorant about the ways in which noise pollution could be controlled. Very few scientists are aware of the problem audits control. Masses are still ignorant of the grave effects produced by noise. In this regard a campaign for wide publicity should be given on TV, radio and newspaper, etc. The Supreme Court in a recent decision has urged the owners of cinema halls and video houses to show environmental slides before the beginning of the movie, in order to disseminate environmental education. The Supreme Court has also ordered all States and Central Government to make circulation/ syllabi for the purpose of introducing environment as a compulsory subject in schools and colleges.
6. Finally, we have seen that in India, in the absence of a specific legislation for control and preservation of noise pollution, one has to seek provisions in various branches of law and regulations which are inadequate. It is high time that Government should pass "Noise Pollution Control Act" to meet specific Indian situations. Apart from such kind of central legislation, there should be a city noise control code for all the major cities of India. Creation of unnecessary noise should be prohibited and should be made punishable (punitive) under law.

Thus it is necessary to take lessons from history where

nourishing civilisations perished not at the hands of any cruel enemy but due to the lack of awareness for maintaining neat and clean environment. The preservation of environment free from noise pollution becomes the concern not only of policy-makers, legislators, judiciary but of all. There is a growing need for proper environment planning and management in the right earnest. The tools in the hands of man are again the same which pollutes the atmosphere and also save it, I.e. science and technology. Ultimately answer has to come from man himself. He who pollutes the environment has to manage it also if he has to survive, otherwise he will suffocate in an environment full of pollutants. Let us be nature's patriots. The life will survive only if the biosphere is safe. The purpose of development should not be to develop things but to develop man.

Planned urbanisation, adequate public transport system, encouraging green belts in cities and adjacent areas, proper sewage treatment system, scientific disposal of wastes and garbage, and all the more public awakening and participation are necessary steps which can prevent all sorts of pollution. Right action at right time is necessary before it is too late.

7

Environmental Laws and Human Rights

> "The execution of laws is more important than the making of them."
>
> —*Thomas Jefferson*

New Agenda for Weary Empire (1989), observed that there are two great dangers of this age; one that humanity will wreak mass destruction with weapons and, two that we destroy our civilization nonviolently—but nonetheless relentlessly—through our indifference and our lack of adequate response to the global environmental crisis that is so clearly advancing upon us. While the threat of an all-out nuclear war has subsised in the past some time, George Kennan's observation that we are destroying our planet Earth nonviolently and inexorably appears to be true indeed.

For the first time in human history we have the power to alter our planet irreversibly on a global scale. Depletion of the ozone layer, increasing rate of species extinction, disposal of toxic and nuclear wastes in vulnerable areas of the planet, loss of forests and of arable soil at alarming rates, are just a few of the many global environmental changes that will affect the survival of the present and future generations on earth.

Notably, in the recent past, technology in chemical and nuclear power plants, land storage and transportation of toxic and radioactive nuclear materials and wastes to and from those plants, have wrought havoc and mass casualties in several parts of the world. Each person's life today depends on the fallible performance of many more individuals. Carelessness by welders and maintenance workers in a reactor or chemical plant could kill thousands and hundreds of thousands not only in the plant and its vicinity but within a considerable distance away from it.

Similarly, a truck driver or railroad engineer could cause mass deaths and environmental destruction through carelessness while transporting dangerous wastes from a chemical plant or a nuclear reactor. Thus the chances of serious accidents that could wipe out human, animal and plant life and destroy the environment will steadily increase unless urgent collaborative action is taken on a global level.

Again storage and disposal of toxic wastes and radioactive materials can also present serious problems. Toxic wastes are indiscriminately stored or disposed of, or clandestinely dumped in nations waterways or in oceans, or exported to countries In the process of development.

Nuclear wastes are improperly stored and escape from tanks from time to time, and thus far no satisfactory solution has been found for permanently storing the hundreds of thousands of tons of spent fuel stored on reactor sites. What can we, as people, as communities, as nations, do to protect ourselves and our environment, our generation and future generations, against what appears to be an inexorable move toward self-destruction? Here we will first discuss the role of national and international law, especially human rights law in the control and prevention of technological disasters and the protection and preservation of the global environment.

Major Issues: From Regulation to Criminalization of Harmful Conduct to Charter of Human Rights

Regulation at the National Level

Global environmental consciousness has had a short history. The first casualties and human sufferings as a direct result of environmental pollution were reported from the industrial areas of Japan in the late 1950's when scores of people died of mercury and cadmium poisoning and hundreds contracted the painful mina mata and itai diseases. Until then there was no real concern for water and air pollution. Mina mata and itai diseases galvanized Japan and the western world to immediate legislative action resulting the "first generation" environmental protection laws of the late 1960's and early 1970's and the establishment of environment ministries in government administration.

These laws and government ministries attempted to control "persistent pollution problems" through regulation of industrial discharges, requiring treatment of effluents at source, filtration and chlorination of drinking water supplies, improvement of technology and other means. The discovery that some industrial chemicals and wastes were highly toxic and resulted in cancer and other malingnant diseases in laboratory animals tested in large scale experiments as well as in human beings exposed to them, especially in workers and in people living near industrial sites, and in miscarriages and birth defects, and in laboratory animals tested in large scale experiments, led to the "second generation" environmental laws of the mid-1970's and early 1980's.

These laws sought to impose controls on the production of certain chemicals and pesticides known to be highly toxic in laboratory animal tests, banned a small number that epidemiological tests proved to be cancerous to human beings and required testing of new chemicals prior to their production, importation, use or marketing. They established stringent standards and controls on the storage, transportation and disposal of hazardous materials and substances. The last decades also witnessed the "third generation" laws.

Protection of the environment and humankind has proceeded from, two directions. The first was a response to human tradegies following several industrial disasters in the 1970's and 1980's, from Flixborough and Seveso in Europe to Bhopal (India) and Mexico. Examples of these laws are the European Community's Seveso Directive issued to member states in 1982. India's Environment (Amendment) Act of 1985 (better known as the Bhopal Act and the emergency preparedness laws of the United States (Superfund Amendment and Reauthorization Act of 1986), of France (1987) and Canada (1988).

Environmental disasters themselves, where no human life has been lost or endangered are addressed by other laws such as the Canadian Environmental Protection Act of 1988. Industrialized countries are however experiencing a great deal of problems in instituting measures to protect the environment itself against cataclysmic disasters posed by the depletion of the ozone layer, global warming, deforestation and desesrtification, the poisoning of the air, water and soil, and the perturbation of the ecosystem by large-scale hydro- electric, irrigation lands water diversion projects. Many of these problems are global in scope and can only be tackled at the global level.

Progress in legal development has been extremely slow but recent international consensus which has produced the Ozone Convention and Protocols on the limitation of CFE production, the Basel Convention and on the control of transboundary movements of hazaradous wastes and their disposal, and the amendment of the London Dumping Convention which will ban dumping of Industrial waste at sea by 1995, bear well for the future cooperation. Nationally, some countries and the European Community have enacted laws, land others such as Canada have tabled draft laws before Parliament, requiring environmental impact assessment of major projects, but a global environmental impact assessment convention is a distant possibility. Worker protection laws have not had the same success as environmental protection laws.

Amendments to Labour Codes have met stiff opposition in industrialized countries and regulatory agencies have been unable to establish exposure levels and standards for more than a handful of the most toxic substances found in the workplace. There is a greater tendency to insist on cost-benefit analysis before exposure levels are reduced than is the case for accident reduction.

A modest "cancer policy" proposed by the Occupational Safety and Health Organisation of the United States in the early 1980's failed to overcome legal challenge in the U.S. Supreme Court. On the other hand, despite regulatory failures, workers in industries with strong unions fared better in achieving protection through the bargaining process. Several industrial countries have enacted laws requiring worker representation on working conditions committees, protecting "whistle blowers" to alert regulatory agencies to dangerous conditions in the workplace, giving workers the right to refuse unsafe work, and in some cases even giving them wage protection when a plant is ordered closed for noncompliance with environmental or labour laws.

In non-unionized sectors where unions are not strong workers, have not fared as well. Furthermore, a very few countries have ratified, or even signed, some of important I.L.O. conventions on worker health and safety, such as the Occupational Cancer Convention, the Occupational Health and Safety Convention and the Asbestos Convention that were concluded in the last fifteen years.

Criminalization of Environmentally Dangerous Conduct

As such while regulation has been the principal mechanism adopted by states to control, reduce, prevent or deter environmentally dangerous conduct, it suffers from several limitations. Regulation requires constant vigilance and action by government agencies. It depends on the establishment of standards, prescriptions, use of specific processes and equipment, and the monitoring to compliance.

Regulatory agencies have to depend on data concerning the safety of chemicals, processes and technology, as well as on economic and technical feasibility. Finally, regulatory agencies have to deploy inspectors and enforcement officers to ensure that the regulations are complied with.

It is clear from the "State of the Environment" reports published by the United Nations Environment Programme in the past few years, that regulation has failed to improve environmental conditions in industrialized countries; at best it has only arrested rapid deterioration of the quality of the environment and the state of human health.

Notably, many countries have therefore introduced a second control mechanism to deter and punish the most serious misconduct of regulated enterprisers. They have enacted special criminal laws or amended their penal codes by creating new crimes in order to prosecute the most flagrant offenders through the criminal justice system.

Japanese law for the punishment of crimes relating to environmental pollution illustrates the first approach. Penal codes of the Federal Republic of Germany, German Democratic Republic (as it then was), Hungary, Portugal, Spain, Austria, and Brazil, illustrate the second approach.

Some of these penal codes do not create what are called "generic" crimes but are, to use the Germany expression, "akzessorisch" to regulatory laws since the commission of the code crimes is dependent on the violation of specific requirements or prohibitions established by regulatory laws. Countries that do not legally distinguish between regulatory offences and crimes, such as Canada and the United States, have created new regulatory crimes dealing with flagrant violations and provided stiff punishment of fines and prison sentences to violators.

Because criminal offences, whether incorporated in the penal codes, in special laws, or in the regulatory laws themselves, depend on standards and prescriptions, the utility of the criminal sanction has been only to stigmatize flagrant misconduct and to punish offenders with heavier penalties. Criminal sanctions however lose their effectiveness when dealing with organizational offenders "which have no soul to be damned or body to be kicked", unless a culpable human agent can be identified for the misdeeds of the organization.

In civil law countries corporations cannot commit crimes and hence it is necessary to identify the human agent that is responsible for the crime. Corporations under common law are prosecutable under criminal law but the extreme penalty of incarceration can only be imposed on culpable human agents. Identification of human agents under both legal systems is problematic specially in large, diffused corporations and hence corporations often get away with monetary penalties however heavy they might be.

Some courts, especially in the United States, have resorted to the theory of vicarious liability for breach of supervisory responsibility and have held high ranking corporate officials guilty of offences committed by those under their control, sending them to prison. See for example, U.S. v Dotterweich (1943 and U.S. v Park (1975) decided by the U.S. Supreme Court. The human agents in small, closely held corporations, cannot escape criminal liability as easily as those in large corporations. Proposals for generic crimes of endangerment and crimes against the environment have been made in some countries by law reform bodies, for example in Canada and by intergovernmental organizations such as the Council of Europe.

Criminalisation of serious environmental misconduct has also been barged by the Eight United Nations Congress on the Prevention of Crime and the Treatment of Offenders in a resolution adopted at the Assembly in Havana, Cuba, in September 1990.

Environment Protection Through the Constitution

Remarkably, over the last few decades the concept of a right to a safe and clean environment has emerged as one on the basic rights in a number of States. Some States consider that the traditional fundamental rights to life, corporeal integrity and health which they recognize in their constitutions, include the Federal Republic of Germany, Italy, the Netherlands, Hungary and Belgium, France,

Austria and Japan. Some countries have introduced environmental protection in the constitution by restricting some other human right, e.g. the right of ownership or right of free enterprise. Examples are Malta. Seychelles, Bahamas, and the constitutions of several former British colonies. On the other hand, the duty imposed on the State by the Turkish Constitution (Article 10) to ensure that every citizen leads a healthy life both physically and mentally, is qualified by the words "to the extent that economic development and financial resources permit". When specific environmental rights have been incorporated in national constitutions they are commonly framed as guidelines, putting the State under an obligation to provide general protection of the environment.

An example of this is article 24 of the Greek Constitution which says:

> "Protection of the natural and cultural environment is a duty of the State. The State is bound to adopt special protection or prohibitive measures for the preservation of the environment".

Article 24.7 of the Swiss Constitution contains a similar guideline. A number of other countries, e.g. India, Sri Lanka, China and some Eastern European countries have similar guidelines in their Constitutions.

Thus, very few countries however recognise environmental protection as a fundamental right. Where it is so recognised the duty to protect the environment is cast both on the State and its citizens. For example, article 45 of the Spanish Constitution states:

> "Everyone has the right to enjoy an environment suitable for the development of the person as the duty to preserve it.'

Article 66 (Portugal), article 39 (Cuba), article 262 (Brazil) and article 15 (the Philippines) are other examples of constitutions which contain a fundamental right to a safe, clean, ecologically balanced (or some other quality of) environment.

Spain and Portugal complete the declaration of a human right with appropriate obligations on the State to undertake measures to implement that right. Article 45 of the Spanish Constitution, for instance says:

> For those who violate (the right) penal administrative sanctions .... shall be established and they (the voilators) shall be obliged to repair the damege caused".

The constitutions of a few U.S. states also contain a right to clean and safe environment: e.g. Msassachusestts (art. XLIX), North Carolina (sec. 5) Pennsylvania (art. 1, sec. 27), South Carolina (art. XI, sec. 1) and Virginia (art. s 1).

The elevation of the right to a safe and clean environment the status of a human right advances the rights of citizens to the highest possible level of protection that can be secured through law. With such constitutional guarantees, citizens can call the State and its organs to account for their acts and ommissions, before the judicial system. All national laws will be measured from the constitutional yardstick making it possible for citizens to obtain remedies against the State.

Similarly, other "debtors" of the right namely individuals and organizations, and even citizens, on whom duties are imposed by the Constitution to protect and preserve the environment, can also to be called to account by the beneficiaries of the right before their judicial system, independently of sectoral laws that may permit their activities.

This feature distinguishes the right to environment from the traditional rights which are framed as restraints on the State and State Organs and not as rights of a citizen or community against other citizens or communities.

The constitution thus be a very effective mechanism to obtain redress in the form of injunctive relief, compensation and restoration of the environment, and other remedies against the State and the other "debtors" of the right. The legal right appears to be complete when, as in the Spanish and Portuguese constitutions, positive duties are imposed on the State to undertake measures to implement the right.

Because of the far reaching ramifications of the human right to a safe and healthy environment, there is understandable resistance to entrenching such a right in the Constitution. It would place a powerful weapon in the hands of defenders of the environment in their battle against proponents of natural resource development.

It could override other individual rights in the Constitution unless there is a provision for its limitation. It would force governments to listen attentively to the demands of citizens and take important resource decisions only after meaningful public participation and in accordance with the majority wishes if not consensus. An important limitation of the extent of constitutional right to a healthy land safe environment is the precision of the definition of that right.

What is the environment? Does it comprises only the constitutive elements of air, water, soil, space, etc., or it the ecosystem in its entirety including the interrelationships of these constitutive elements? How healthy is "healthy", "clean", "sound", etc., the terms that are often used to describe the right? Consensus is very difficult on these questions.

A second, and perhaps more serious limitation is the failure of the lawmakers to consider the claims of future generations to an environment which is at least as healthy and safe as that enjoyed by the current generation, and which will provide them with adequate and undepleted natural resources to survive. Most of the constitutions referred to above almost entirely ignore inter-generational equity between present and future generations.

Regulation by Virtue of Conventions and Treaties

It is significant to note that International law has made most astonishing progress in the last few decades in the realm of environmental protection. Bilateral treaties between countries sharing one or more natural resources, regional agreements among several countries which are part of the same ecosystem, and multilateral conventions bind in countries sharing the global commons, have significantly impingned upon the principle of state sovereignty that has been the hallmark of international relations in the past.

Regional, supra-national, inter- governmental and international organizations are now playing a far greater role than in the past in the development of policies, guidelines, recommendations, regulations and conventions, and in the administration of these instruments.

There are a large number of international arrangements addressing environmental relations between and among States. In the field of resource conservation we find several fisheries conservation conventions, migratory species convanent and endangered species conventions. In the field of ocean protection, the most significant convention that currently binds 64 countries is the nearly twenty-year old London Dumping Convention which controls or prohibits the dumping by ships of certain hazardous industrial wastes into the sea. This convention is administered by the International Maritime Organization which also administers a few other pollution of the sea and liability conventions.

Another important multilateral convention which contains a specific part on the protection of the marine environment is the United Nations Convention on the Law of the Seas of 1982. Several regional seas protection convention have also been concluded in various regional areas under the auspices of the United Nations Environmental Programme. In the field of atmospheric protection, the two important multilateral instruments concluded in the last dozen years and the ECE Convention on the Long Range Transport of Air Pollutants (1979) and the Convention on the Protection of the Ozone layer (1985). Both are framework conventions requiring specific protocols to reduce the emissions of pollutants covered in them, e.g. sulphur dioxide, nitrogen oxide, cholorofluorocarbons, etc.

Notably, the control of transboundary movement and disposal of hazardous wastes not covered, by the London Dumping Convention is now the subject of the Basel Convention concluded in March 1989. The United Nations Environment Programme has provided the interim secretariat for this convention under the International Register of Potentially Toxic Chemicals (IRPTC).

In the field of toxic chemical and pesticides, four important organizations are playing crucial role in development of national and international law: the Organisation for Economic Cooperation and Development (OECD), the IRPTC, the International Programme on Chemical Safety (IPCS), and the Food and Agriculture Organization (FAO).

Under the asupices of the OECD the 24 member countries have taken several recommendations, guidelines, etc. in the area of testing, good laboratory practices, information exchange, pre-market notification, manufacture, export of toxic chemicals, etc. These instruments have various legal implications to member State obligations: decisions are generally binding on them, recommendations are not legally binding but are generally adopted, and guidelines and recommended practices are generally adhered to because of the commonality of interest of the members who represent most of the producing and user states. The IRPTC is an international clearing house established by the United Nations Environmental Programme (UNEP) for scientific, technical, legal and regulatory information for assessment and control of chemical hazards. It administer the Provisional Notification Scheme on banned or severely restricted chemicals, which was adopted by UNEP in 1984 and disseminates the information provided to by UNEP member under the Scheme to other members.

While the IPCS is a cooperative venture of the WHO, ILO and UNEP to evaluate the effects of chemicals on health and the environment. It carries our risk assessments on dozens of chemicals and pesticides and publishes environmental health criteria documents on them.

Similarly, the FAO has been very active in the area of food safety and pesticide. Its Codex Alimentarius Commission develops, establishes and publishes, food safety standards for a large number of food products, including food additives, pesticides and other cotaminants. FAO's International Code on the Distribution of the use of pesticides, adopted in 1985, is designed to promote the safe and efficient use of pesticides and to prevent their misuse. It contains duties and responsibilities, and a voluntary code of conduct for all public and private entities engaged in or effecting the distribution and use of pesticides such as governments, industry, wholesale and retail vendors and users of these.

In fact, protection of workers' health is the main concern of the International labour Organization (ILO), a specialized agency of the United Nations. The ILOs contribution is mainly through standard-setting in the form of conventions and recommendations,, and guidelines in the form of Codes of Practice etc. The major conventions developed the ILO in the worker health protection area include the Occupational Cancer Convention of 1974, the Occupational Safety and Health Convention of 1981 and the Asbestos Convention of 1986. Following the Bhopal tragedy which killed hundreds of workers and nearby residents, the ILO Instituted the Occupational Safety and Health Hazard Alert System.

Further, World Bank and the United Nations Centre on Transnational Corporations (UNCTC) are two other important institutions concerned with health and safety and the environment. The World Bank has been the promoter of large-scale development projects in countries of the industrially developing world, and is able to use its financing leverage to ensure balanced economic development and the use of best available technology that will protect workers and minimize environmental harm. In 1985, the Bank published a Manual of Industrial Hazard Assessment Techniques for the prevention of major accidents. The Bank can play a crucial role by using its financial clout to ensure that the assisted countries protect the human rights of people forcibly displaced by major hydroelectric, irrigation and water diversion projects which it finances.

In the similar vein, the U.N. Centre on Transnational Corporations has recently finalized a draft code of conduct for Multinational Corporations which imposes duties and obligations on these corporations. It has also published a book on Requirements for Information Disclosure in Environmental hazards, based upon the lessons learned in industrial environment protection.

Safe and Healthy Environment as a Human Right

The constitutional recognition of a human right to safe and healthy environment by the several countries has had minimal impact on the level of protection actually enjoyed by the most vulnerable individuals and groups in those countries. While human rights law and an activist judiciary can assist in the achievement of such protection, there is a far greater need for organizational, informational, and financial assistance to help them achieve meaningful standards.

This need is most acute when the people exposed to dangers, and those forcibly evicted by development projects, live in poverty or on the fringes of society and are thus most vulnerable to human rights abuses. Human rights abuses can occur not only in the hands of governments by their actions and inactions but also from acts of foreign governments and their instrumantalities. When foreign governments permit their citizens and corporations to export hazardous substances and technologies, without adequate controls and safeguards and without considering the capacity of the receiving countries to cope with unintended and unforeseeable consequences of operation, they are in fact allowing "perverse development" to occur in those nations.

In fact, a human right to a healthy and safe environment at the international level is almost non-existent. There are as yet no treaties providing for such a right although thereafter a number of international declarations and proposals which urge governments ~~to~~ recognize and establish the right. Very recently, the rights of future generations has also received some type of recognition. The human right to a healthy and safe environment was first recognized in the United Nations Universal Declaration of Human Rights of 1948. Article 25 of the Declaration recognizes the right to a standard of living adequate for the health and well-being of an individual. Article 3 can also form the basis of a safe and healthy environment since it recognizes the right to life.

The right to a healthy and ecologically balanced environment was reiterated in the 1972 Stockholm Declaration at the United Nations Conference on Environment. Paragraph I of the preamble to the declaration states: "Man has the fundamental right to freedom, equality that permits a life of dignity and well-being, and he bears a solemn responsibility to protect and improve the environment of present and future generations".

and paragraph I of the article 25 states:

> "Everyone has the right to a standard of living adequate for the health and well-being of himself and his family."
>
> Neither of these two Declarations are legally binding, but they can provide precedents for the future.

The International Covenant on Economic, Social, and Cultural Rights, which is an International document legally binding on its signatories does however contain a provision which refers to a healthy and safe environment. Paragraph I of article 12 states:

> "States parties (to the Covenant) recognize the right of everyone to the highest attainable standards of physical and mental health, and agree to take steps to achieve the full realization of this right, including

the improvement of all aspects of environmental and industrial hygiene".

According to Jemm-Uuri Uibopuu, this article has raised the matter of human environment "from the exclusive sphere of state competence and made it a matter of international concern in International Covenant law." He adds that not only environmentally harmful transnational acts are considered but also those acts occur within the boundaries of one State. The International Covenant also contains another article which guarantees the right to life, "which shall not be arbitrarily deprived of Article 6(1). While it is arguable that the right should at least be interpreted to impose an obligation of states to protect citizens against all kinds of threats to human life, including ecological dangers, Dr Uibopuu considers it as referring to the protection of the mere existence of the human being without pointing at any quality of life and well-being.

The right to a healthy and safe environment has also been subject of two recent international documents. The first is the Final report of the World Expert Group on Environmental law, which was established by the Brundtland Commission in 1985. The fundamental right to an environment adequate for the health and well-being of all human beings is contained in Article 1 of the proposed General Principles concerning natural resources and environmental interferences. An article 2 of the General Principles call upon States to ensure that the environment and natural sources and conserved and used for the benefit of present and future generations.

The second document is the Declaration of the Hague signed by heads of 22 governments in March 1989. The declaration, which was issued in the context of global climate change states:

> "The right to live is the right from which all the other rights stem. Guranteeing the right is the paramount duty of those in charge of all States through the world.
>
> "Because the problems of warming the global atmosphere and of deterioration of the ozone layer are planet-wide in scope, solutions can only be developed on a global level. The remedies to be sought involve not only the fundamental duty to preserve the ecosystem but also the right to live in a viable global environment and the consequent duty of the nations vis-a-vis present and the future generations to do all that can be done to preserve the quality of the atmosphere".

Also, a pronouncement on human rights and the environment is contained in Resolution 199040 of 6th March 1990 which was adopted at the 52nd meeting of the commission on Human Rights of the United Nations Economic and Social Council. The resolution calls upon (governments of all countries of the world) to do their utmost to assist in implementing the right to life through the adoption of appropriate measures at both the national and international levels:

> "calls upon all States, appropriate United Nations bodies specialized agencies and inter-governmental and non-governmental organizations concerned to take the necessary measures to ensure that the results of scientific and technological progress ... are used for the benefit of mankind and for promoting and encouraging universal respect for human rights and fundamental freedoms."

Agenda for Action

The survey of national and international laws indicated that a human right to a safe and healthy environment has not fared well in national and international covenants. State charters of rights and freedoms direct too much attention to the rights of individuals vis-a-vis the State when in reality society is composed of many groups and the State is not often the most determinative factor.

In fact the rights of an individual against the State are also rights of the individual over other persons and over nature. The very exercise of liberty creates space for domination by some over others and there is no assurance that human rights against the State will not be employed so as cause lawful harm to others. Courts of some countries have acted in innovative ways to protect citizens against these kinds of lawful harm, by holding the State responsible not only for its acts of commission but also omission and by creating a duty upon the State.

They have imposed an obligation on the State to regulate and to enforce regulations. Many individuals and organizations, public and private, national as well as transnational, conduct activities that profoundly affect the health and well-being of people who do not have the political or economic power to protect themselves from the effects of those activities. Many "development wrongs can be identified : high risk activities which endanger the life and health of workers in a plant or of people living in its vicinity, discrimination against disadvantaged groups or their exclusion and marginalization in the development process, environmental harms, exploitation resources without any concern for the needs of future generation or even the future needs of the present generation, abject poverty, and various cultural harms. A growing number of people believe that these individuals and institutions must be held accountable to more rigorous standards with a view to protecting the V human rights of people affected by their activities. Their accountability can be advanced by strengthening national law, especially human rights law, applying international Instruments, and furthering rights of people at all levels to participate so that they themselves can shape the ends and means of development.

Transboundary pollution and the causation of transboundary effects remain serious problems, both on a routine day-to-day basis (e.g. acid rain, global warming, ozone layer depletion) as well as on a catastrophic disaster basis (e.g., Chernobyl, Basel and Valdez). The absence of international standards and the multiplicity of national standards further exacerbate the problems.

Capacity to inflict harms may be truly global but liability and remedies are not. The concept of multinational enterprise liability remains inadequately developed under international law. Many of the actions of the present generation may impose serious environmental burdens on future generations. By exhausting resources each generation narrows the range of resources available to its descendants.

The way we dispose of waste affects the quality of the environment that we pass on to our children and their children. If we reap short-term benefits from cheap disposal of wastes in the air, land, fresh water, seas or even outer space, we may pass on staggering costs on to future generations which must either try to cleanse the disposal sites or abandon them altogether.

We also degrade environmental quality for future generations by destroying forests, soils and watersheds which provide essential environmental services. Forest, for example can prevent soil erosion, facilitate the recharge of the water table, harbour wildlife, maintain soil productivity and buffer climate etc.

Agenda for National and International Action

Law persons concerned with the current state of environmental human rights should consider the need to evolve new jurisprudence and new clusters of human rights to protect the present and future, generations, especially those people most vulnerable to "development wrongs" identified earlier: the poor, the disadvantaged and workers in hazardous enterprises.

National Action

Some of the dye components of those rights might include:

I. Right of empowerment and people participation, including

(i) the expansion of the right to life, liberty and security of the person enshrined in national constitutions to include the right to a safe and healthy environment, and

(ii) an environmental bill of rights some of the key elements of which would be: (a) a right to know and freedom of information, (b) participatory risk assessments and acceptance, and (c) new compensation mechanisms as part of incentives rather than remedies, presented to communities at risk in advance of a decisions to site a particular project to enable them to make risk acceptance decisions for themselves.

- equitable distribution of both risks and benefits.
- new concept of accountability whereby wielders of public power are made accountable to the people affected by it and to the people at large.

- new system of compensation for victims, separating the fact-finding components from the compensation award components.
- right of monitoring the implementation of safety measures and enforcement of regulations.

II. A charter of rights for workers in hazardous industries which would include:

1. freedom of choice and freedom from economic coercion, right to refuse unsafe work, protection of "whistle blowers" who alert authorities to unsafe working conditions, protection of wages when a plant is shut down for correcting those conditions;
2. right to participate in management decisions that affect safety standards and workplace conditions;
3. right to know the hazardous or toxic nature of substance that will be coming into contact at the plant;
4. right to accurate accident reporting and health reporting systems; and
5. right to qualified and safe co-workers.

III. A charter of rights for future generations which would;

1. recognise the right of future generations to a right to environmental quality comparable to that received and enjoyed by previous generations;
2. impose a duty of stewardship on each generation conserve and use the environment and natural resources for the benefit of itself and future generations and to prevent irreversible and irreparable harm to life on earth;

3. make each generation accountable to the next generation at the same time that it exercises its right to use the natural environment for its own benefit. This accountability might be secured through:

   (a) representation of future generations in decision-making (non-governmental organizations could act as trustees; they could perhaps fill this role better than State representatives);

   (b) sustainable development of renewal resources and monitoring of the diversity and quality of the environment;

   (c) inter-generational trust funds financed by global user fees or tools, such as fees on land disposal of hazardous waste, "severance tax" on extraction on non-renewal resources, fee for using the global commons, e.g. "carbon tax" on motor vehicles and on products or services that contaminate the earth's air, atmosphere and waterways.

International Action

There is a clear need for developing international norms, standards and human rights, to move away from national law to international responsibility. International action should include:

(i) the development of the concept of multinational enterprise liability of encompassing both civil and criminal liability

(ii) a comprehensive system of regulation at national and international levels governing all aspects of chemical and nuclear hazards, such as the export, import, manufacture, use, transport, storage, and waste disposal and including research, especially on toxic effects and treatment, plant design, safety measures and operating procedures, workers' rights, plant siting and pollution;

(iii) the development of international systems for hazard assessment;

(iv) the improvement of the international information systems regarding the toxic effects of chemicals by strengthening and extending the mandate of the International Register of Potentially Toxic Chemicals:

(v) creation of an international disaster relief system;

(vi) development of a more effective international justice system for the victims, for example special rules under international private law on jurisdiction, forum standing, discovery, representation, due process, interim orders, recognition and enforcement of foreign judgments, and a new international dispute processing system for victims of environmental disasters, and

(vii) the expansion of human rights in the International Covenant on Economic, Social and Cultural Rights to include specifically the right to a safe and healthy environment using the special appeal process provided in it for references to the Human Rights Commission.

Recently in Dec. 2009 a summit is held in Copenhagen which thought could not reach unanimous conclusion, decided to cut carbon emission at individual level by the participant.

Land and Water Rights

When we speak of our Environment we refer to land, sea, air on our planetary home Earth as well as its outer space. The right of individuals to be guaranteed to pure decent and healthy environment can be conceived as an emerging human right, inspite of the notion that the individual not a subject of international law, and Article 2(7) of the United Nations Charter giving inherent right of Sovereignty to the State.

The grave calamities at Bhopal, Chernobyle and Basal illustrate the fundamental recognition of the right of individuals namely the right to live which is protected under Article 6 of the Declaration of Human Rights (1948) which is also protected under Article 6 of the International Covenant on Civil and political rights which is Js Cogens and cannot be derogated by State even during the period of emergency.

This historic Stockholm Conference of 1972 represents the first major effort to solve the global problem of protection and improvement of the environment by international agreements on universal levels as far as possible. It was the beginning of the recognition of basic human right to environment.

As per 'the principle 21 of the 1972 Conference the natural resources of the earth including the air, water, land, flora and fauna and especially representative samples of natural ecosystem must, be safeguarded for the benefit of present and future generation. It established a balance between the sovereign rights to use one's own natural resources and the duty to avoid damage to other. As the mankind approaches the twenty-first century there is a need to strengthen the protection and enforcement of human rights guarantee of environment being destroyed by poor agriculture, and increasing population pressure and rapid industrialisation and urbanisation have led to the necessity of additional living space and concrete jungles, replaces forests at the alarming rate of 300 thousand acres an hour and almost 1/3 of the world forests have already been cleared.

Hazards at Global Level

The poor irrigation and over use of grasslands, droughts, famines and soil erosion, salivation, alkalization and water logging is increasing our deserts by 40 thousand miles a year.

Industrialization is considered as the sine quo non of the progress especially in the developing countries.

Eight million hectares of land every year is being converted for urbanization, road construction, living spaces and industrialization every year.

The danger of beautiful earth, blue-sky and starful nights turning into a perpetual planet of darkness and death is becoming more real than ever before because of pollution. Not all the global environmentation problems were foreseeable in 1972. The World Commission on Environment and Development in its report in 1987 says: Our common future recognised that "all human beings have the fundamental rights to an environment adequate for their health and well being and further lays down the requirement to provide compensation for activities on breach of international environmental obligation".

The environment that sustain life on Earth is in peril. Human actions are responsible. We must learn to live in greater harmony with our planet than we have so far done.

There is a deepening environmental crisis in all the parts of the world. In the developed and affluent countries because of industrialization and in the under-developed countries because of poverty etc.

We are depleting our natural resources at rates which threaten their continuation for future generations. Our forest are being cut down ten times more quickly than they were being replaced. Our soil are being destroyed by erosion.

Those who by their acts, endanger the environment on which life depends clearly are committing an act of aggression against human rights, and the right to life itself.

The US President John. F. Kennedy on September 20,1963 addressing the UN General Assembly said "never before has man had such capacity to control his own environment to end thirst and hunger, to conquer poverty and disease; to banish illiteracy and massive human misery. We have the powers to make this the best generation of mankind or to take it the last."

It is becoming increasingly clear that the sources and causes of pollution are far more different, complex and inter-related and the effects of pollution more widespread, cumulative and chronic than hitherto believed. Industrialized countries generate about 90% of the world's hazardous wastes. Their energy consumption is polluting the biosphere.

If the present trends continue, the people of the planet face a gloomy prospect. The man is unable to breath even pure air, drink pure water and eat pure food. The prospect of mankind slowly dying under the influence of man-made negating environment is more real than the thermo conflict apparently.

Adopting Strategy

Infact, there is a need of a global strategy for protection of the environment and the rational use of natural resources and solving the problems which include climate changes, transboundary movements of pollutants acid rain that deplete ozone layer, the pollution of the world oceans, soil degradation, desertification deforestation, the disappearance of species of the flora and fauna, the destruction of polar region, the recycling of decontaminated and disposal of hazardous wastes, genetic effect, etc. There is hardly any state in the world now whose top leaders have not focused their attention to environment issues, treating the issue at par with other pressing problems. Ecology does not respect national boundaries.

No country can develop in isolation from others, hence new orientation is required in international relations for development. There is not a need even to redefine development. Human laws must be reformulated to keep human activities in harmony with the unchanging and universal laws of nature. There is a need to an integrated international approach for need to V development and protection of environment. Anthony D Mato, writing in *American Journal of International Law* vol. 84(i) January 1990 said: 'we hold the earth is trust for further generation at the same time we are beneficiaries entitled to use and benefit from it' (p. 190).

Prof. J.G. Strake has also admitted that "the international law of development has not yet reached the stage where it can be set down as a substantial body on binding rule conferring specific rights upon developing states and imposing duties on developed countries".

For the developing countries it is not only the production but also in social transformation equal distribution and uplift of the poor and the backward and fighting the illiteracy, disease, poverty and unemployment and to achieve all this, there is a need of new social and economic order. The scientific research and technology for development leading to self-sufficiency and fulfillment of national requirements within available natural resources.

Indian Scheme

India has taken a lead in giving a constitutional status to Environment. The Constitution (42nd amendment act 1976 adds an article 48A to the part IV of the Directive Principles as under:

> "The state shall endeavor to protect and improve the environment to safeguard the forests and wild life of the country.

Article 51A of the Fundamental Duties prescribes "it shall be the duty of every citizen of India to protect and improve that natural environment including forests, lakes, rivers and wild life and to have compassion for living creators."

The Water (Prevention and Control of Pollution) Act 1974 followed by Air (Prevention and Control of Pollution) Act 1988; and last but not the least Environment Protection Act 1986, go a long way to prove the awareness of the problem by the people of India and their search for the solutions. Pre-Stockholm conference legislations are also worth mentioning here in order to show the concern for environment in India. The Indian Penal Code of 1860 makes it punishable offence of polluting water of any public reservoir stream.

Similarly Northern India Canal and Drainage Act 1873 as amended and Indian Fisheries Act 1897, Indian Forests Act 1927 and Orissa River Pollution Prevention Act 1954 are some of the pioneer steps taken towards pollution control in India.

No doubt, there is a tradition of considering the water of river "Ganga" as 'holy and "pure", but the purity of Ganga and many other rivers including Jamuna, Krishna Kaveri, Narmada, etc have been complete erred from the very source itself due to industrial discharge, sewage, fertilizer, cement and the pesticides factories on their banks, besides sewage disposal. The environment protection policy has to be formulated and implemented through a comprehensive legislature covering all aspects of environment rather than peace meal-solution piecemeal problems.

It is remarkable that the law has a positive role to keep our ecological balance by introducing "Judicial restraints as well as judicial activism".

There is need to transform the interpretation of right to life in Article 21 of the Constitution into a meaningful right to healthy life and healthy environment with human dignity.

The post-Maneka Gandhi case (AIR 1978 SC 597) puts an obligation on the state to take the affirmative steps not only to protect life but also to protect environment for life. In a developing country like ours, there is lack of technological resources and proper planning and enforcement of pollution controlling policies. The Supreme Court of India has been in public interest litigants expounding the need for proper clean, hygienic and safe environment from pollution.

As a healthy development, in recent years lawyers have joined ecologists in defending the rights of people to have pure water, air. They have been debating the rights and duties of the people, state, and industrialists to safeguard and preserve the nature and environment so that the natural resources are conserved and degradation of the environment is reduced to a minimum. We have to achieve the equity within the present generation and also future generation. Protecting the environment is paramount duty of the states and its citizen.

The duty to protect the environment cannot be let of the customary laws but should be enforced through bilateral and universal treaties at international level and municipal laws with 'strict liability clause at the national level'. The threat to our ecosystem is so real that it is turning our fertile areas into virtual deserts and forcing our hapless people to become environmental refugees. With almost 23 of our population living in sub-human conditions poverty must be regarded as the biggest source of Pollution.

The former Prime Minister of India, Indira Gandhi while addressing UN Conference on Human Rights on 6th June 1972 said "Environmental problems of developing countries are not the side effects of excessive industrialization but reflect the inadequacy of development-are not poverty and need the greatest polluters".

What Can be Done More?

There is need of World Conference on Environment on Development in order to check wanton exploitation of nature. Tree plantation and tree farming can remove the pollution and poverty from the earth. We should have one point programme to reduce our needs. God has created enough to fulfill the needs of man on this earth but there shall never be enough to fulfill the greeds of man.

Our holy scriptures say: a man who plants tree is far more superior than a person producing 100 sons. Should we not like to become superior persons by planting a tree rather than increasing the population. The whole world will become our beautiful garden. We shall get rid of the pollution of air and water and soil erosion. There shall be enough food, oxygen and drinking water. In this world of consumerism we are consuming the whole earth, thereby depriving ourselves and our future generation the natural resources on which the whole life depends.

The environment world-wide is in crisis. Although this message has for many years been propagated by environmental movements and scientists, it has also been accepted in recent years by the establishment, including the world's political leaders who met at the United Nations Conference on Environment and Development (Earth Summit), Brazil, in June, 1992; Every citizen has the right to live in an environment worthy of human existence. This is natural, as the concept of Human Rights is also a historical legal category.

Its 200 year-old history proves that its individual components have been enriched in the course of social development. The sporadic industrial development and the environmental protection coupled with it have also upset the rights to life and health in other aspects. With the beginning of research activities in the field of the environmental law with some confusion at the beginning concerning the proper place of the environmental protection in the legal system.

In the modern society, with the complexity of intensifying socio-economic relationship there is a growing tension between the structure of circumstances to be regulated and the structure of the legal system. This has an influence on the structure of the legal system on the formation of the interdisciplinary instruments in other words, the quasi branches of law.

Hence, following the initial exploratory work, specialists

in environmental protection law are of the opinion that it does not form part of any kind of an already existing branch, of law, but it is such a complex, inter-disciplinary law province, in which administrative law, land law, civil 'law, International law, and criminal law are equally involved.

Such inter-disciplinary field of environmental protection law, necessarily takes it closer to the 'concept and concern of Human Rights. This interdisciplinary attitude of environmental laws creates several consequences directly effecting the Human Rights.

On the other hand, the role of International cooperation in environmental protection has also assumed tremendous significance. Air and water are not discriminated with respect to social systems of societies and countries.

Consequently, international cooperation, the activity of international organisations, their recommendations, the regulations in the form of mutual international agreements play an ever increasing role in the regulation of environmental protection.

Legal science has, basically, two attitudes as to whether they take national or international law as a basis. These solutions are easy either to defend or to a attack, though the basic problem is, that it is not only national and international but universal human relations that really matter.

Anyhow, it is an accepted rule in the international system of environmental protection - as a minimal right or obligation of countries i.e., a state should act in a way so as to not cause any harm to other countries. This principle has been accepted both in Stockholm and in Helsinki and more recently in Copenhagen.

It is well settled that in the absence of efficient and effective legal measures the declaration of the right to clean environment, as a human right would remain merely a phrase, and of little utility. Since law deals with the protection of the human environment from pollution, harm, damage and detriments in three dimensions -Legislation, Law observance and the application of law - and it also classifies them. In each dimensions and in each legal solution action must be based on actual social conditions with circumspection to modern technical data.

As we live in society where law plays a necessary role in environmental protection, though an indirect one especially in the elimination of detrimental factors in the environment or it can be a link of a chain in reaching a particular result. The system of legal norms - including the coercive power of the State is the entity of permissive, limiting and prohibitive measures.

It aims at the prevention of environmental harm, the control of environmental factors and the application of sanctions against those who violate the permissive limiting or prohibitive measures. It is namely the starting point of law that the different degrees of pollution are not independent of the human mind and will, but there are circumstances which are recognisable and surmountable.

Therefore, for the lawyer, it is never a particular factory, machine, energy, active substance that means file final source of the pollution but always the human activity hidden behind it. Similarly, the final target of the pollution is also man itself. Law as a means of environmental control, combines a definite social requirement with scientific recognition together with a respect for the economic capacity of society, and on the basis of these it forms a system of legal norms in environmental control.

In fact, India is an original signatory to the solemn Declaration of the United Nations Conference held in Stockholm on Human Environment in year 1972. This Declaration while acting as an eye-opener to mankind and the people in the world on tile essential and imperative need to protect the environment makes an interesting reading and hence it is felt a part of it is worth recapitulation. Man is both creature and moulder of his environment which gives him physical sustenance and affords him the opportunity for intellectual, moral, social and the spiritual growth. In the long and tortuous revolution of the human race on this planet, a stage has been reached when, though the rapid, acceleration of science and technology, man has acquired the power to transform his environment in countless ways and on an unprecedented scale. Both aspects of man's environment the natural and the man-made, are essential to his well-being and to the enjoyment of basic human rights. Even the right to life itself.

Further, India is one of the very few countries in the world which has provided for Constitutional safeguards for the protection and preservation of the environment. Article 48-A provides protection and improvement of environment and safeguarding of forests and wild life. The State shall endeavour to protect and improve the environment and to safeguard the forests and wild life of the country. Furthermore, duties of the citizen towards environment are contained in Article 51-A, "It shall be the duty of every citizen of India to protect and improve the natural environment including forests, rivers and wildlife and to have compassion of living creatures". Apart from Constitutional safeguards there is a contingent of about more than 300 Central and State legislations which have a direct or an indirect bearing on the environment. However, their efficacy is questionable. Most of the legislations dealing directly with Environment Protection are known more for their violation than implementation in India.

The critical analysis of the implementation of Water (Prevention & Control of Pollution) Act, 1974 and the Air (Prevention and Pollution Act), 1981 is in itself glaring illustration of total failure of these legislations in attaining the aim and object of their enactment by the Parliament. There are legislations with respect to the conservation of forests and Wild Life, namely, the Indian Forest Act, 1927, The Forest (Conservation) Act, 1980 (Amended 1992). The Wild Life Protection Act, 1972. However, notwithstanding the powers and the celebrated objectives for which the enactment has come into force, there is a wide gap between what the legislation seeks to provide for and what actually is practised?

Another important legislation is the Factories Act, 1948. This Act is applicable to all the factories and under different chapters, it provides for certain measures with respect to the health of workers, industrial safety welfare measures and a new chapter dealing with hazardous processes. The authority concerned with implementing this legislation is the State Government. This is a well thought out legislation, covering all aspects of the activities of the factory. However, the Bhopal Gas Tragedy and the recent decisions of the Apex Court of the country concerning the functioning and management of factories and their hazardous effect in polluting the environment are contrary to the spirit of this legislation. The aforesaid discussion makes it clear that in our country there is a plethora of legislations ranging from Acts on Health, Insecticides, Explosive substances, Drugs and Cosmetics, Food Adulteration, Inflammable Substances, Arms, Poisons, Preservation of Forests, Wildlife to Protect Water and Air Pollution, etc.

Yet the problem of environmental degradation is becoming grave complex and unmanageable. There is no dearth of environmental protection but we need a firm hand to implement them.

After a brief review of the environmental legislations The Tiwari Committee of 1980, noted some major shortcomings which can be summarized as below:

1. Many of these laws are outdated.
2. They lack statements of explicit policy objectives.
3. They are mutually inconsistent.
4. They lack adequate provisions for helping the implementing machinery.
5. There is no procedure for reviewing the efficacy of the laws.

On the basis of its analysis the Committee suggested the following main recommendations:

1. Comprehensive review and reformation of some Central and State Acts (such as the Insecticides Act, 1968, The Water (Preservation and Control of Pollution) Act, 1974; and the Indian Forest Act, 1927).
2. New legislation for areas of action not covered by the present laws.
3. The introduction of Environment Protection in the concurrent list of Seventh Schedule.

"Once again, while not denying that many new areas of human activity need to be covered by legislation, one must seriously question the wisdom behind the strategy of merely demanding more and harsher laws to solve the problems. Is this suggestion based on a sound theory, or is it backed by solid empirical evidence that it will work, if anything, all our past experience only point to the contrary. They point to the fact that what is wrong with our laws is not harming of less or weak laws, but harming a wrong type of laws which are often neither effective nor conducive in the attaining of just goals".

Secondly, the disillusionment of the citizens about the sincerity of the Government to protect the environment has increased due to doubts about the objectivity, integrity and efficiency of the agencies entrusted with the implementation of environmental regulations and pollution control laws. Finally, it may be kept in mind that no law howsoever well intentioned and conceived, can succeed in attaining its objectives unless it is understood and supported by the people for whom it is meant. This in short, means public participation both at the stage of formulation as well as the stage of implementation. In a democracy, legislation has to be of the people, by the people and for the people.

Several countries in Europe, recognise the Right to a Clean, "man-worthy" environment as a basic human right. In the Western World great efforts are being made to maintain the environment. The United Nations is very much conscious of the world-wide problem of maintaining the Environment safe for human being. The United Nations Conference on the Human Environment held at Stockholm in June, 1972, paved way for the passing of a resolution by the United Nations General Assembly on December 15, 1972, emphasising the need of active co-operation among the Status in the field of Human Environment. The Resolution had designated June 5 as the World Environment Day and has urged Government and organisations in the United Nations system to undertake on that day every year worldwide activities reaffirming their concern for the preservation and enhancement of the environment.

It is with this objective in mind that the limitations and potentials of environmental protection laws in India need to be thoroughly examined, appropriately amended, properly understood and firmly and rightly implemented.

It is not enough to aspire for integral and upto-date codes. This will partly solve the problem of the quantum of laws. But the problem of their quality remains. The phrase 'quality" of laws refers to many aspects. The questions it invites attention to are: Does the law communicate the message to the addressee? Does it clearly define rights, privileges, powers and immunities of various agencies of the Government and the individual organisations? Is its management of sanctions - both positive and negative (that as rewards incentives and punishment) - adequate to attain the purposes of law? Are the jobs of confining and structuring of adjudication are settlement of disputes efficient and fair? What possibilities for administrative deviance and unjustified harassment does the legislation create, inadvertantly or otherwise. These questions and others are important not just from the stand point of sociology of law. They are quite crucial from technocratic perspectives of natural resource management.

Finally, it becomes evidently clear that for practical observance of the legal measures for the declaration of the right to clean environment as a 'human right' we need effective and efficient legal environmental laws. Such legal measures are to be codified as basic human rights, while its real implementation can be secured only by a broad spectrum of legal political and social norms.

Legal Principles for Environmental Protection and Sustainable Development Adopted by the WCED Experts Group on Environmental Law

General Principles, Rights, and Responsibilities

Fundamental Human Right

1. All human beings have the fundamental right to an environment adequate for their health and well-being.

Inter-Generational Equity

2. States shall conserve and use the environment and natural resources for the benefit of present and future generations.

Conservation and Sustainable Use

3. States shall maintain ecosystems and ecological processes essential for the functioning of the biosphere, shall preserve biological diversity, and shall observe the principle of optimum sustainable yield in the use of living natural resources and ecosystems.

Environmental Standards and Monitoring

4. States shall establish adequate environmental protection standards and monitor changes in and publish relevant data on environmental quality and resource use.

Prior Environmental Assessments

5. States shall make or require prior environmental assessments of proposed activities which may significantly affect the environment or use of a natural resource.

Prior Notification, Access, and Due Process

6. States shall inform in a timely manner all persons likely to be significantly affected by a planned activity and to grant them equal access and due process in administrative and judicial proceedings. Sustainable Development and Assistance.

7. States shall ensure that conservation is treated as an integral part of the planning and implementation of development activities and provide assistance to other States, especially to developing countries, in support of environmental protection and sustainable development.

General Obligation to Cooperate

8. States shall cooperate in good faith with other States in implementing the preceding rights and obligations.

Principles, Rights, and Obligations Concerning Transboundary Natural Resources and Environmental Interferences

Reasonable and Equitable Use

9. States shall use transboundary natural resources in a reasonable and equitable manner.

Prevention and Abatement

10. States shall prevent or abate any transboundary environmental interference which could cause or causes significant harm (but subject to certain exceptions provided for in Art. 11 and Art. 12 below).

Strict Liability

11. States shall take all reasonable precautionary measures to limit the risk when carrying out or permitting certain dangerous but beneficial activities

and shall ensure that compensation is provided should substantial transboundary harm occur even when the activities were not known to be harmful at the time they were undertaken.

Prior Agreements When Prevention Costs Greatly Exceed Harm

12. States shall enter into negotiations with the affected State on the equitable conditions under which the activity could be carried out when planning to carry out or permit activities causing transboundary harm which is substantial but far less than the cost of prevention, (if no agreement can be reached, see Art. 22.)

Non-Discrimination

13. States shall apply as a minimum at least the same standards for environmental conduct and impacts regarding transboundary natural resources and environmental interferences as are applied domestically (i.e., do not do to others what you would not do to your own citizens).

General Obligation to Co-operate on Transboundary Environmental Problems

14. States shall co-operate in good faith with other States to achieve optimal use of transboundary natural resources and effective prevention or abatement of transboundary environmental interferences.

Exchange of Information

15. States of origin shall provide timely and relevant information to the other concerned States regarding

transboundary natural resources or environmental interferences.

Prior Assessment and Notification

16. States shall provide prior and timely notification and relevant information to the other concerned States and shall make or require an environmental assessment of planned activities which may have significant transboundary effects.

Prior Consultations

17. States of origin shall consult at an early stage and in good faith with other concerned States regarding existing or potential transboundary interferences with their use of a natural resource or the environment.

Arrangements for Environmental Assessment and Protection

18. States shall co-operate with the concerned States in monitoring, scientific research and standard setting, regarding transboundary natural resources and environmental interferences.

Emergency Situations

19. State shall develop contingency plans regarding emergency situations likely to cause transboundary environmental interferences and shall promptly warn, provide relevant information to and co-operate with concerned States when emergencies occur.

Equal Access and Treatment

20. States shall grant equal access, due process and equal treatment in administrative and judicial proceedings to all persons who are or may be affected

by transboundary interferences with their use of a natural resource or the environment.

State Responsibility

21. States shall cease activities which breach an international obligation regarding the environment and provide compensation for the harm caused.

Peaceful Settlement of Disputes

22. States shall settle environmental disputes by peaceful means. If mutual agreement on a solution or on other dispute settlement arrangements is not reached within 18 months, the dispute shall be submitted to conciliation and, if unresolved, thereafter to arbitration or judicial settlement at the request of any of the concerned States.

Now we come to Indian scene. The following description is based on the write-up a famous Indian jurist, which deals with the matter in right and more emphatic manner.

Environmental Pollution and Legislative Solution*

The Third World has no tomorrow if it is fatally infactuated with the deadly dope of gargantuan industrialization and imitation projects without the healing hope of anti-pollution strategies successfully adopted by socialistic countries. Paul Harrison has put the point pithily:

'One of the most distasteful aspects of the Western model of development has been its appalling arrogance, in the face both of nature and of traditional cultures. The rape of the earth has been central to the past progress of Western industrialism: the plunder and waste of irreplaceable

* We are grateful to the original contributor for subscribing his learned views.

resources, the poisoning or destruction of fragile ecosystems, the belief that man can mould nature with no need to take account of nature's reactions to such a violation. Disdain for traditional cultures and social systems was a fundamental characteristic of all modern European empires, and this attitude spread to the new ruling elites who had adopted Western ways. As a result one development project after another triggered irreversible changes in local habitats. And the whole development process did not build on the foundations of traditional society but fell on it like a nuclear bomb and blew it apart.

> ....eco-development is no more than a sensible precondition of sustainable development. It asks no more than that we should live today with tomorrow in mind, that we do not snatch a momentary prosperity for ourselves at the expense of the very survival of our children.
>
> Third World leaders frequently refer in their pronouncements to national values and aspirations and to the uniqueness of their local cultures; but these ultimate ends are generally ignored in practice. With a few notable exceptions, development has been a piecemeal, pragmatic endeavour, aimed for the most part at the multiplication of material wealth. Basic values such as justice, cooperation, charity and respect for nature have not only been subordinated to that, but are often trampled underfoot in the gold rush'.
>
> —*Paul Harrison*

A Soviet pamphlet tells as some tragic facts:

> 'Man is facing an ecological crisis of his own making. For centuries he thought little about the consequences of the pressures mankind was placing upon nature.

> Today the by-products of man's activity can be compared to the results of global geological processes.

It has been estimated that every year 16,000 million tons of oxygen are burned up and that between 21,000 and 24,000 million tons of carbon dioxide are discharged into the atmosphere. The pollution rivers, lakes, seas and oceans has reached a dangerous level, with about 50 per cent of all the pollution of the biosphere being caused by the United States, which has less than six per cent of the world's population.

The Austrian scientist and Nobel Laureate's Civilized *Man's Eight Deadly Sins* emphasized that the harmful effects of technological expansion could be substantially reduced if the commercial attitude to nature could be replaced by more rational activity. To achieve this, writes Lorenz, it is necessary to uproot the prevalent psychological attitude of have it now. As long as mankind continues to exploit nature in a blind and barbarous way, it will be constantly threatened by ecological ruin.

Does this blood-chilling forecast apply to the whole of mankind, to all countries?

Likewise Prof. Suresh Srivastava writes:

> 'A recent survey reveals that "every day millions of gallons of trade wastes and effluents are discharged into the rivers, streams, lake and sea etc." Indiscriminate water pollution is a problem all over the world but is now acute in densely populated industrial cities, India is no exception to this.

Air pollution has further added to the intensity and extent of the problem. Like water pollution, every year millions of tons of gaseous and p articulate pollutants are injected into the atmosphere both through natural processes and as a direct result of human activity. Scientists have pointed out that the earth's atmosphere cannot absorb unlimited amount of pollutant materials without undergoing changes which may be of an adverse nature with respect to human welfare.

Noise is another problem of industrialization and extensive use of machinery. Continuous noise seriously affects structural and functional integrity of the ear. Broadly speaking the more intense the noise the greater is the hearing loss; the degree of hearing loss and its position in the tonal spectrum is related to the frequency value noise. The duration of the period to which the subject is exposed to the noise influences the hearing loss.

One reads everyday about the degradation of the environment, rape of the earth, traumatic subversion of the eco-system, poisoning of air, water and food and technological plunder of the resources of Nature. I may perhaps prove this industrial robbery with murderous impact on mankind by citing from the Soviet publication:

> The pollution of air, water and soil, and consequently the contamination of food with chemicals in the economically advanced countries has led scientists to believe that the bacteriological era has come to an end and the new, chemical era has set in.

The pollution of soil and water reservoirs is also increasing. This creates a situation where no more trees, shrubs or even grass grow in large cities.

The damage, and the consequent loss inflicted upon farming and the timber and fishing industries are truly tremendous. That is why environmental hygiene is a very urgent problem. Its solution would save nature, while failure could mean the destruction of the natural environment.

Measures to protect the environment have assumed particularly massive dimensions during the past decades, which can be explained by the growing rate of nature consumption due to the rapid progress of technology.

In connection to with air pollution, Yuri Siuyakov says:

> 'Let's take, as an example, the effect industrial and household aerosols have on the behaviour of our atmosphere. It has been established that in humid air the aerosols stimulate the formation of fogs which, in turn, intensify the concentration and, consequently, the toxicity of additives in the atmosphere. Formed under specific conditions, smog is very dangerous, indeed. The city over which it forms becomes shrouded in a cloud of its own waste matter. . . . the scientific and technological revolution makes it impossible to completely eliminate the dangerous effect of man's economic activity on the atmosphere.

As such the deeper we delve into the wealth of the ocean the greater the ecological damage we inflict. The problem of the protection of water resources against pollution with oil and heavy metals, organic chrolic compounds, radioactive and other toxic substances is a serious one throughout the world, particularly in countries with long coast lines. Water pollution leads to the death of plankton and lowers the intensity of gas metabolism between the world ocean and the atmosphere. Fresh water resources pose many other problems. And although there is a lot of water on our planet, fresh water constitutes a mere two per cent. Most of the fresh water is conserved in the glaciers. And it is precisely fresh water that industry and agriculture increasingly require.

A tremendous quantity of fresh water is used for

irrigation: growing one ton of wheat requires 1,500 tons of water. The growth of cities and the development of industry changed the amount and composition of effluent, and created the most acute problem of our century, the problem of clean fresh water. In one degree or another this problem faces all countries. Every year man increases pressure on the land which remains to be the principal source of food.

To increase land fertility, we plough deeper and more frequently. This leads to the destruction of soil structure and its pulverization, The soil becomes increasingly susceptible to wind and water erosion.

Marine pollution is an imminent oceanic disaster, even as world-wide deforestation is a Himalayan blunder driving towards desertification, reduced rainfall and greenery:

> 'The oceans covering two-thirds of the earth's surface are vital for life on land. They are a treasure of valuable resources; food, minerals, chemicals petroleum, gas, pearls etc. From ancient times the oceans have been used for transportation. Oceans have rightly been called as our last frontiers.

Traditionally, allover the world seas have been used as a dumping ground for wastes. Earlier, wastes included mainly domestic sewage carried by rivers, agricultural wastes etc. But, today modernization of society has been accompanied by pollution of the seas with industrial waste, deliberate, operational and accidental discharges of shipborne petroleum and poisonous gases, disposal of radio-active wastes from uses of nuclear energy, uses of the oceans for military and defence purposes. These pollutants are destructive of marine life and other vital resources in the seas. They snap the food chains on which marine life survives and reduce the availability of oxygen for marine life. Pollution poisons fish and other species consumed by man. Minamata Disease caused by mercury poisoning of fish in Japan in the mid sixties of the last century is an example. Recently Indian shrimps have been found contaminated with cholera gems. Most States are coastal.

In fact, half of the biological productivity of the world's oceans occurs in the coastal regions of the sea. The volume of coastal waters, though only a minimum part of the total ocean, is yet an important consideration with regard to marine pollution, since coastal regions are especially vulnerable to pollution. Much of the waste is discharged directly or indirectly within the coastal waters. Sixty percent of the world's largest cities are located within fifty kilometers of the sea, and half of the world population lives within 200 miles of the seacoast. Commercial fishing in coastal waters is an important industry.

The aesthetic characteristics of the coasts also attract people for recreation. It is a tourist region with significant tourist business. Therefore, it has been recognised that marine environment and all the living organisms which it supports are of vital importance to humanity, and all people, particularly the coastal States, have an understandable interest in assuring that this environment is so managed that its quality and resources are not impaired.

More recently, with atomic weaponry and other radioactive substances entering the human theatre, the exposure of the eco-systems to great dangers is grave indeed. We see the escalating potential of industrialization to imperil, even, irreversibly, the basics of ecological harmony and the happy balance in Nature.

Humanity is indivisible. So is peace and security. Civilized man's uncivilized aggression on Nature has made us realize that, like humanity, ecology is indivisible. Peace and Salvation on earth are possible only if we are conscious of ecological ethics. We must forbid industrialist privateers and individualist techno-buccaneers from piracy of Nature molesting the flora and fauna and polluting the biosphere.

The fact that ecological questions have become one indivisible global problem is a salient feature of our time. The problem has become international, giving a new dimension to international relations. Ecological security has become a firmly established term in modern international parlance, since it reflects the desire of every nation to protect the environment and the health of people which may be affected by the anti-ecological actions of other countries. Today with the scope of man's economic activities so wide that they affect the natural environment of a number of countries, it has become necessary to develop ecological ethics.

Man's only known home, Planet Earth, is under homicidal threat through pollution of environment:

> "The source of life is largely oxygen and this human planet of ours will perish if by pollution we destroy that fountain. And yet, our oceans which are basic to much needed oxygen are being poisoned continually by civilisation which is assiduously digging its own grave. Law is a means to an end. The primary end is assurance of human survival and so the first charge on the legal system is elimination of pollution of those vital sources without which life on earth may prove impossible".

Proximate international peril has quickend the United Nations' attention to the calamity and the industrial nations show concern Joint discussions, regional co-operation and collaboration between countries in the area of conservation and anti pollution research are now a reality. National

Constitutions are showing ecological awareness. For example, the Soviet Constitution (Article 18) says:

> 'In the interests of the present and future generations, the necessary steps are taken in the USSR to protect and make scientific, rational use of the land and its mineral and water resources, and the plant and animal kingdoms, to preserve the purity of air and water, ensure reproduction of natural wealth, and improve the human environment".

Let us get the Indian back-drop to the ecological crisis and environmental pollution. Our guilt began when we assassinated Gandhi, not physically but developmentally. This humanist, moral phenomenon, Gandhi, was against inhuman industrialization and prescribed processes of national regeration which entitled him to be called by Schumacher 'the greatest economist of the century'. Had we followed Gandhi the Socialist Republic India would not have reincarnated as the Great Indian Moral-Material Desert. Industrialize or perish was the slogan with which we began our journey from Independence; and now, Industrialize and perish is the unwritten epitaph which a freebooters' economy, socialist manifestos notwithstanding, has immorally managed.

The dual dimensions of pollution - moral and maternal - should not be missed. India is no exception. What with trans-national corporations with unlimited exploitative appetites, infra-national industrialists with imitative tactics and money power at various levels wooing political power and white collar, we have unconscionable ecocides who seduce politicians, smog the vision of governments, lubricate the wheels of the bureaucracy and propagandize pollution as a necessary evil for the salvation of the nation.

Notably, the Indian ecological facts are agonising. Pollution of Nature can be, often is, of inner space - moral-mental - and of outer space - environmental. If the moral fibre fails and mental perception is blunted - the methods are well-known - seminars on techniques of Operation Clean-up and legislations aping English and American models, may prove to be damp squibs.

So first, we must start and offensive against moral pollution without which scientists will not come out with the truth, ministers will surrender to dubious pressure and monitoring boards will go to innocent sleep and even law courts may decline to do justice to Nature due to guiltless illiteracy and social insensitivity.

A cultural campaign is necessary to rouse the conscience and consciousness of the community to the calamitous after-effects of the degradation of environment and its permanent pauperisation of the people. Sans social consciousness and mass mobilization, the voice of the scientist-sentinel will be stifled by Money Power and the soul of those who matter in the legislative, executive and judicative branches will slump.

It is apt to quote Ralph Nadir has said:

> In no clearer fashion has the corporation held the law at bay than in the latter's paralysis toward the corporate crime wave. Crime statistics almost wholly ignore corporate or business crime; there is no list of the ten most wanted corporations; the law affords no means of regularly collecting" data on corporate crime; and much corporate criminal behaviour (such as pollution) has not been made a crime because of corporate opposition. For example, willful and knowing violations of auto, tire, radiation, and gas pipe-line safety standards are not considered crimes under the

> relevant statutes even if lives are lost as a result. The description of an array of corporate crimes in this forthright book reveals a legal process requiring courage not routine, duty, by officials to enforce the laws against such outrages. The law is much more comfortable sentencing a telephone coin box thief to five years than sentencing a billion-dollar price-fixing executive to six weeks in jail. In one recounting after another, the authors pile up the evidence towards one searing conclusion - that corporate economic product, and environmental crimes dwarf other crimes in damage to health, safety and property, in confiscation or theft of other people's monies, and in control of the agencies which are supposed to stop this crime and fraud. And it all goes on year after year by blue-chip corporate recidivsts.
>
> (Foreword, *America Inc....*)

In India the private sector operators sullies water and air, spoils ocean and sky, spoliates forests and poisons farms and plantations; he kills birds, beasts and fisheries and behaves more irresponsibly by pleading that safety is costly, knowing that life is cheap. The poignant facts is that the public sector too behaves no better. Even the Planning Commission woke up to the seriousness of the problem only around 1965. In our country a little pragmatism and foresight on the part of industries and government would have definitely abated pollution to a great extent.

Until the Fourth Five Year National Economic Plan (1969-70 to 1973-74), the significance of the quality of environment to the aim of bettering the quality of life was not considered. The Plan for the first time took cognizance of air and water pollution, soil erosion, waste of national resources etc. It states that each generation has the obligation to maintain the productive capacity of land, air, water and wildlife so as to leave successors some choice in the creation of a healthy environment.

Principal growth in India is envisaged in the fields of power, petrochemicals, fertilisers, synthetic fibres, metallurgy, sulphuric acid, nitric acid, pharmaceutical industries and various other chemicals and ceramic industries. In addition to industries, the other important source of air pollution in India is the domestic consumption of low grade fuels, resulting in intensely smoky atmosphere. Contribution of the fine dust by the deserts and other open dry fields and the unpaved streets is also considerable.

Auto-exhaust is yet another contribution to the environmental pollution in the cities. Eighty per cent of the Indian automobiles are more than five years old. Because of this and because of improper maintenance they emit large quantities of carbon monoxide, hydro-carbons, oxides of nitrogen and other pollutants.

As such in India, air pollution is caused by industrial emissions, transport emissions, meteoriological conditions and improper land use, without any planning for air pollution control. The industries in India have never considered pollution control equipment as an important component of their plant equipment or social accountability.

Pollution of the environment has many dimensions. Third World countries have problems of mass poverty the abolition of which is the first covenant in our tryst with destiny. To repeat what the West does is simian stupidity because our priorities are different. Urban and agrestic squalour are a major pollutant in India.

Deforestation and denudation have made lush Kerala brown and barren and Andamans an enemy of trees! At this rate, Kashmir may be a naked dawsel before long!

Project physical sanitation is a feasible, not fantastic, scheme provided Sewagram or, Kasturbagram, were made a national model for immediate planned implementation. China, with a billion people in poverty, has changed the humanscape through gladsome communes and disciplined cadres. How clean, neat, loving and yet blending with Nature, without the wealth of the West and the filth of the East! We too can, and that, by law and life together. Water Act and Air Act are ounces of opium so long as the Panchayat Act and Municipalities Act do not act to eradicate the slums and cesspools, the drainless sullied water and generous spread of garbage litter and the animal farm conditions for Indian man's residence! These local bodies are moribund organs, from the angle of providing sanitation and health like functioning drains and public toilets, potable water and smoke-free air, mandatory under the Act. If Law is what law does, not what law says, from the Pollution point of view, Law is dead.

All the meagre resources are spent on salaries and wages with little left for even basic anti-pollution, pro-public health provisions. Bureaucratisation has made these local self-government units local babels of politics with no effective initiative, no meaningful responsibility, no creative opportunity, free from the inhibitive hierarchy of officers and patronising over lordship of ministers. If the people and their representatives do not participate constructively in Operation Clean-up, paper self-govt. alone win exist. So it is that democratisation of local bodies with people's involvement in municipal functions is the only method of activising the environmental mandates. A full-blooded, people-oriented panchayati raj through new legislation is the first requisite. A mass-line, whereby each ward or sub-village will be under the charge of a batch of public-spirited persons, is the second imperative.

The springing up of voluntary agencies working against local pollution can be good catalysts.

India's pollution problems are many. We suffer at once from too much primitivism and too much modernism. Primitivism can be conquered by adopting the Chinese, why Gandhian line. Night soil, garbage and all refuse can be used for manure and biogas.

The potential of biogas is considerable. If all the dung of India's livestock, and all human wastes, were fermented, this would meet the domestic fuel requirements of three out of four families and the savings in manure or fuel could be worth $1 billion a year or more. Other countries are now realising the importance of biogas. China, which does not have India's cultural and social obstacles, uses biogas widely; in 1977 around 5 million plants were in use.'

Chemical fertilisers are doubly injurious and violates Nature. We can ill-afford the cost. After all the hungry have to be fed.

The human bird under the Indian sky gets up weaker than when he pretended to retire. For millions it is an eternal vigil or an eternal trance. It is an indescribably painful state which has to be experienced to be realised.

There is on the face of the earth no other country that has the problem that India has of chronic starvation and slow death - a process of de-humanization. The solution must, therefore, be original. In trying to find it, we must discover the causes of the tremendous tragedy.

This is true of Free India's to-day as of imperial yesterday. Legal status, not voluntary officiousness, must be imparted to peoples- role in pollution control - not only in village and town but in melas and fairs, festivals and pilgrim centres.

The pathology of modern industrialism is the familiar story of the West, with the added misfortune that many of our large industries of the Big Houses have political clout and hardly worry about anti-pollution obligation before and after the Water Act and what not. As a fact, the corporate capacity to outwit the law is not peculiar to pollution control measures but to all economic regulations. This may be called the blue-chip syndrome in Americanised diction.

In India big timber traders spirit away the best wood from government forests, even lobby for hydel projects as that gives an alibi for deforestation.

But now, in nostalgic mood, some Frost with frosted soul, may sing: 'The woods were lovely, green and deep! The birds "have flown, the arborial game has fled, the trees have reincarnated as profits of timber traders and the parabiad soil has bid farewell to Nature and eroded because ecology is not a word in the millionaire's prayers. The process is plunder. The forest department from the political top onto the legal echelons and forest administrators, is the collective agent of desrtification through deforestation. All private forests were nationalised in Kerala and they are now an' orphanage. The story of the tree is the same from the Himalayas to Sahyadri, thanks to the commercial soul of the arboricidal combine. Even the enchantment of the Andamans is now the exasperation of the ecologist: The Law, in letter, forbids but, in action, yawns.

Even the State is reckless and the populous poor of the Third World are a scapegoat in this process. Biogas, available as in Chinese communes and Gandhian programmes out of village refuse, is dismissed as primitive by the Indian elite and forests are used as fuel or exported as finished goods. Paul Harrison sums up one facet:

> 'Finding appropriate solutions to the fuel crisis is becoming an important priority, both in alleviating poverty and in preserving the precarious environment The World's tropical forests are being felled at such a speed that at present rates the forests of the developing countries could well disappear within a century, and more than four-fifths of the wood that is cut is being used for fuel. As population expands, the task of gathering fuel has become increasingly time-consuming and exorbitantly expensive. The resulting deforestation is stripping soil of its cover, accelerating erosion and the spread of deserts. Thus the poor are driven by their sheer poverty to destroy the environment they depend on for a living-they are buying short-term survival by committing long-term suicide'.

The Forest Act, the nationalisation legislation and, in some States, laws to preserve private forests exist. But the law, equal in its majesty, operates obliquely, as the anonymous but acid lines tell:

> 'The law locks up both man and woman.,
> Who steals the goose from off the common,
> But lets the greater felon loose,
> Who steals the common from the goose.'

The answer? An environment Ombudsman, not the creature of a party but the choice of parliament, with immunities of a High Court Judge and powers, judicative, executive and punitive, set by the law with technical and administrative infrastructure. A pre-condition, of course, is societal awareness and culture of reverence for Nature.

The India saga of homicidal environmental misfeasance and ecological lawlessness follows the pattern of the capitalist West although we are in a Socialist Republic with constitutional conservational commands. To sum up, many industries pollute air and water producing disease and death. Many plants have smogged out the sun and many wood-based mills have wiped out rain forests and raped virgin soil. Many aigricultural practices, with pesticides and fertilisers, have inaugurated the era of forbidden fruits and grains. To-day, hungry for profits, lives on Tomorrow of blurred distance. The politician is concerned with elections and the statesman is concerned with generations. The former is in power and makes hay with Big business as abetor. The latter is an endangered species and is dismissed as Utopian. Even our Public Sector, victim of the pathological strategy, of survival through suicide, makes no bones about pollution, ecology and Nature nonsense. From the Ganga and Yamuna to Krishna, many innocent streams are contaminated; holdy mountains, treeless and thirsty, stare at the sun. The pathos and bathos of the State of our natural environment is evidence that we are not socialist. Look at Soviet Legislation and projects for conservation and learn to be sensitised about the social future of our Socialist Republic.

The management of the plant, together with scientist, developed a project for reducing the discharge with special trapping devices, which cost five million roubles for its realization. Now the plant practically does not pollute the environment. And the park which has been laid out nearby symbolizes the peaceful co-existence of chemistry and nature. There are many other examples of this kind.

Economic calculations show that the cost of amelioration and health improvement measures is worthwhile both in terms of social benefit and money. The improvement of the surrounding atmosphere by the management of one of the biggest centres of the chemical industry at Berezniki (Perm region, Russian Federation) cost 260 million roubles. The sharp reduction in the atmospheric pollution made it possible to reduce the sanitary-protective zone from 5 kilometres to 255 KMs. And this saved the nation 250 million roubles, to say nothing of the tremendous advantages of utilizing valuable chemicals which had been discharged into the atmosphere.

What had literally gone up in smoke is now put to good use. Another example is the use of filters at industrial enterprises at the town of Apatity, beyond the Arctic circle at the foot of the khibiny mountains. As the name of the city suggests, the main local industry is the production of apatites (the 'stone of soil fertility') which are shipped out in large quantities. In addition to the rock apatite, there is apatite dust which is collected by special trapping installations. According to specialists, this dust has a very high commercial value.

These tasks are stressed in the decisions of the 25th Congress of the CPSU. The adoption of new technological processes that preclude the discharge of toxic substances is one way of combating atmosphere pollution. But the greatest effect can be achieved only when all these methods are combined -when the composition of the fuels used by factories and transport is improved a more ration distribution of the sources of toxic discharge is made, the area devoted to greenery is expanded, and city planning is improved."

Standing up for Nature is integral to ecological harmony and socialist policy. Man cannot go it alone and must advance in alliance with not in defiance of Nature. This is compulsive, scientific and socialist. So, farsighted Lenin planned Soviet developmental strategics on that basic premise.

"The Soviet Union is the pioneer of environmental protection. In the 1920s nature conservation was elevated to the level of state policy. The desire that the rapid scientific and technical progress should proceed side by side with the rational and thrity development of natural resources without ever harming the environment is a reflection of the general attitude of the Soviet People during the entire period of socialist construction.

From the first decrees of the Soviet Government to the present the fundamental principles of the socialist attitude to the natural environment were established by Lenin who, on Nov. 7,1917, the day of the victory of the socialist revolution in Petrograd, drafted the text of the Decree on Land which was adopted by the Second Congress of Soviets the following day. Since then the entire land, its mineral resources, water, forests and natural wealth became the property of the state, the property of the entire nation. This legislative act inaugurated a new relationship between nature and society.

In 1918 the Fundamental Law on Forests was adopted requiring all local organs of power to make sure that the forest reserves of the country would not be depleted.

On July 20, 1920, the Decree on Hunting established game preserves and hunting rounds. A sweeping programme was also initiated for saving animals which were on the verge of extinction. To carry out that programme the Government allocated considerable sums of money. Many Soviet schools of higher learning began to train hunting specialists, animal breeders and ichthyologists.

With Indo-Soviet collaboration at a high level, we may look at their experience also and not be lost in American meditation.

If human welfare, over the long run, is the foundation of our Republic which is the happy sanctuary of 700 million (now over 1000 million) people, surely we must have a dynamic environmental policy to the accompaniment of protective legislation and effective implementation.

Luckily, the United Nations is waking up to the world peril. The Helsinki Conference showed concern for environmental protection in the Final Act and defined areas of co-operation. Since then the UN Economic Commission for Europe, UNESCO, WHO and other bodies have initiated research and education on the subject. More than all, the nidus and impetus for environmental pollution legislation and measures for protection is the United Nations Declaration on the Human Environment, Stockholm, 1972.

Having considered the need for a common outlook and for common principles to inspire and guide the Peoples of the world in the preservation and enhancement of the human environment

1. Proclaims that: Man is both creature and moulder of his environment, which gives him physical sustenance and affords him the opportunity for intellectual, moral, social and spiritual growth. In the long and tortuous evolution of the human race on this planet a stage has been reached when, through the rapid acceleration of science and technology, man has acquired the power to transform his environment in countless ways and on an unprecedented scale. Both aspects of man's environment, the natural and the man-made, are essential to his well being and to the enjoyment of basic human rights - even the right to life itself.

   The protection and improvement of the human

environment is a major issue which affects the well-being of peoples and economic development throughout the world, it is the urgent desire of the peoples of the whole world and the duty of all Governments.

2. (ex. 2) The natural resources of the earth, including the air, water, land, flora and fauna and especially representative samples of natural ecosystems, must be safeguarded for the benefit of present and future generations through careful planning or management, as appropriate.
3. (ex. 3) The capacity of the earth to produce vital renewable resources must be maintained and wherever practicable restored or improved.
4. (new) Man has a special responsibility to safeguard and wisely manage the heritage of wildlife and its habitat which are now gravely imperilled by a combination of adverse factors. Nature conservation including wildlife must therefore receive importance in planning for economic development.
5. (ex. 4) The non-renewable resources of the earth must be employed in such a way as to guard against the danger of their future exhaustion and to ensure that benefits from such employment are shared by all mankind.
6. (ex. 14) Appropriate national institutions must be entrusted with the task of planning, managing or controlling the environmental resources of States with the view to enhancing environmental quality.
7. (ex. 15) Science and technology as part of their contribution to economic and social developmental must be applied to the identification, avoidance and control of environmental risks and the solution

of environmental problems and for the common good of mankind.

The Socialist States, including China, have taken many measures of conservation with social sensitivity and humanist concern. Our jurists, saturated with Anglo-American legal learning, are not fully aware of the happenings in the crimson world. However, a perusal of any Indian seminar paper or legal article will flatteringly refer to American law on the subject of Ralph Nader and others of his like fight for an environmental fair deal against the subversive Corporate Sector which de facto owns or controls the United States thro' devious ways. Let us also battle like that, with law as weapon.

Be that as it may, broad thro' rudimentary, sufficient thro' somnolescent, legal control of pollution was available even in British India. We have earlier mentioned how most noxious conditions could be interdicted, arrested or corrected by alert and affirmative steps under municipal regulations. And the court could enforce these duties which local bodies owe to the local people.

Juristic archaeologists, willing to dig into legislative debris will discover that the Britisher wanted Indians to keep away from pollution.

The earliest Act on the statute book concerning control of water pollution in India is the Shore Nuisance (Bombay and Kolaba) Act 1853 which declared it lawful for the Collector of Land Revenue, Bombay to give notice to an offending party, requiring the removal of any nuisance anywhere below high-water mark in the Bombay Harbour. If the nuisance was not removed or abated within one month of notice, the Collector was authorised to get it removed or abated.

Another old statute is the Orient Gas Company Act of 1857 that comprehensively attempts to thwart water pollution such as that which concerned the company. The Oriental Gas Company was registered in England as a joint stock company and was later incorporated in India for the purpose of manufacture and supply of gas for lighting the town of Calcutta and its neighbourhood. It was found expedient to give this company powers and facilities to enable, them to carry out their undertaking of lighting with gas the city of Calcutta under statutory provisions. Section 15 of the Act relating to water pollution reads as follows:

> Penalty for causing water to be corrupted. Daily penalty during the contimance of the offence.

Later, Section 17 of the same Act provided the following penalty if water be fouled by gas.

The Serais Act of 1867 enjoined upon a keeper of a serai or an inn to keep a certain quality of water fit for consumption by 'persons and animals using it' to the satisfaction of the District Magistrate or his nominees. Failure for maintaining the standard entailed a liability of rupees twenty.

The Northern India Canal and Drainage Act was passed in 1873 which listed certain offences under the Act contained in section 70. Sub-clause (3) of Section 70 provides that any interference with or alteration in the flow of water in any river or stream, so as to endanger, damage or render less useful any canal or drainage-work would be an offence. Sub-clause (5) later provides that whoever corrupts or fouls the water of any canal so as to render it less fit for the purposes for which it is ordinarily used would be an offence. The penalty for the breach of this provision would be imprisonment not exceeding one month or a fine not exceeding fifty rupees or both.

Section 8 of the Obstruction in Fairways Act of 1881 empowered the Central Government to make Rules to regulate or prohibit the throwing of rubbish in any fairway leading to a port causing or likely to give rise to a bank or shoal.

Water pollution by oil has been regulated by the Indians Ports Act, 1908.

The Inland Steam Vessels Act, 1917 lays down a mandatory rule of fresh water for the use of passengers.

Section 26 (i) of the Indian Forest Act, 1927 makes it punishable if any person, who in contravention of the rules made by the State Government, poisons water of a forest area.

There are other micro provisions and minor enactments banning water pollution even smoke nuisance e.g. Bengal and Smoke Nuisance Act, 1905.

The criminal law catalogue is not complete without reference to two major enactments - the Criminal Pro. Code 1898, (now 1973) and the Indian Penal Code, 1860.

The Indian Penal Code, 1872 : Section 277 of the Code says: "Whoever voluntarily corrupts or fouls the water of any public spring or reservoir, so as to render it less fit for the purpose for which it is ordinarily used, shall be punished with imprisonment of either description for a term which may extend to three months, or with fine which may tend to five hundred rupees, or with both. This provision is extremely limited in scope. It applies only to a public spring or reservoir. It has been held that the term public spring does not include a public river or water flowing in a continuous stream in a river bed. On the same analogy a canal or any other running water will not be covered by the term. A well, however, is a spring. A reservoir may be natural or artificial. A lake will be covered by the term reservoir.

There is section 269 which prescribes punishment for negligent act likely to spread infection of disease dangerous to life. It says:

> Whoever unlawfully or negligently does any act which is, and which he knows or has reason to believe to be, likely to spread the infection of any disease dangerous to life, shall be punished with imprisonment of either description for a term which may extend to six months, or with fine, or with both.

Pollution of waters other than springs and reservoirs will be covered by section 290 of the Code. It provides: Whoever commits a public nuisance in any case not otherwise punishable by the Code, shall be punished with fine which may extend to two hundred rupees. Since pollution of a public stream is a public nuisance, a pollution of the same could be punished under the section. However, the quantum of punishment is extremely small. Public nuisance is defined in section 268. The idea behind public nuisance is causing any common injury, danger or annoyance to the public.

Another section of the Indian Penal Code relevant to pollution of water is section 426 which deals with mischief. It says: 'Whoever commits mischief shall be punished with imprisonment of either description for a term which may extend to three months, or with fine, or both'. Mischief is defined in section 425 as causing destruction of any property or any such change in any property or in the situation thereof as destroys or diminishes its, value or utility, or affects it injuriously.

Hardly, any case law on pollution has occured under the Indian Penal Code which goes to show its infirmity in dealing with the problems of pollution. Recently the directors and manager of the Gwalior Silk Mills were prosecuted under Sections 268, 269, 277, 288 and 290 of the I.P.C. for polluting the Chambal River by putting certain waste into it from the factory. It was held by the Madhya Pradesh High Court in Babulal V. Ghanshyamdas Birla that the directors and other officials could not be prosecuted unless a specific act or omission was attributed to them or there was an overt act done by them and there was mens era on their part.

The Procedure Code, likewise, vests powers to stop a nuisance or other injury in two broadly worded provisions viz. S Secs. 133 and 144. The power is pervasive and can instantly prevent any noxious activity or nuisance provided alertly sensitized imagination and socially generous interpretation play upon the provisions, the authorities are fearlessly vigilant to initiate public interest action against pollution invasion. The law limped and rarely helped because the activist awareness was absent. Municipal paralysis in minimal anti-pollution measures is endemic to India because local self-government is often nominal and, at best, vocal. Blaming the British but remaining bureaucratic or politically elitist is no solution. Local Laws and the people are still distant neighbours.

Local Boards Acts if democratically implemented, Town Nuisance Acts and the Police Acts, if promptly and punitively used, the Criminal Procedure Code and the Indian Penal Code if socially activated and proscriptively popularised, almost all nuisances, widely defined to include pollution on land, contamination of water, noise aggression and noxious discharges into the atmosphere in short, biosphere molestation, could be interdicted. But environmental protection is not the executive's cup of tea nor the judiciary street. After the Ratlam Municipality case the vistas are wide and public interest litigation is a punitive panacea. Today, people have become pachydermic that not only the big industries but the pilgrim centres and political melas are also guilty of pollution. Look at every 'temple and mosque with loud-speakers and amplifiers, religiously disturbing the people's peace and sleep with midnight bhajans and noisy discourses. See how every festival and marriage occasion assaults ordinary citizens with furious gramophone music amplified to boost the locals blood pressure by deprivation of quiet. What tongue can adequately tell how the speeding auto-monsters - the trucks and buses - blow into innocent's environment wicked black fumes and blast the neighbourhood with deafening sound waves- The Indian road syndrome is unique in the world. Induration of people's and officials' pollution consciousness makes pollutional chaos a way of life in India. Law is too timed to halt the polities of pollution. Here again, China is a contrast of quite and clean milieu now.

Taking an over-view of the British Indian legislation for supression of pollution of land, water and air, we find that, through the Victorian era and after there have been fragmentary statutory exercises consistent with the limited problems and perspections of those times, although judged by the giant strides of industrial pollution and advance of environmentology today, those laws may look primitive and naive. However, had those enactments in their full creative potential been put into operation, substantial success could have been achieved. But the laws did not keep their promises because the implementation was entrusted to the beaurocracy and, Public interest litigation was an unknown quantity. Of course when over-zealous officers wanted to put an end to public nuisance, in its wide connotation, - they did succeed. After an pollution of land, water and air noise pollution, marine pollution and other environmental menaces could be ' brought under the rubric or nuisance provided the enforcing agency and the judicial authority were ready to adopt a generous interpretation and crusading spirit.

There is no gain-saying in the fact that having regard to the modern dimensions and difficult patterns of pollution more expertise in the statute and enforcement authorities is a desideratum. Even modern Municipal statutes are more specialised. As example the Delhi Municipal Corporation Act, 1957 may be looked into. The Act empowers the commissioner to make an order restraining the use of water from any well, tank or other source of supply not vested in the Corporation when it is so polluted as to be prejudicial to health of the people. Restrictions have been placed on construction of latrine or cesspool near a well, tank, water pipe or cistern. Section 241 requires that no person shall throw, empty, or turn into any municipal drain or into any drain communicating with a municipal drain.

(a) any matter likely to injure the drain or to interfere with the free flow of its contents or to affect prejudicially the treatment and disposal of its contents; or

(b) any chemical, refuse or waste stream, or any liquid of a temperature higher than forty five degrees centigrade, being refuse or stream, or a liquid which when so heated is, either alone or in combination with the contents of the drain, dangerous, or the cause of a nuisance, or prejudicial to health; or

(c) any dangerous petroleum.

Section 242 prohibits a person to discharge into a municipal drain any trade effluent from any trade premises except in accordance with bye-laws made by the Corporation.

Section 250 provides that no sewage shall be discharged into any water course until it has been so treated as not to affect prejudicially the purity and quality of the water into which it is discharged.

Section 354 provides that all fifth and polluted and obnoxious matter accumulating upon the premises of a person is to be collected and removed to specified places in the prescribed manner. Section 356 has specific provisions with regard to accumulation and the removal of rubbish belonging to factories and workshops. The Commissioner has been given power to ask the factory owners to remove the rubbish, filth and other polluted or obnoxious matter to specified places in the prescribed manner. The Commissioner may himself do so and charge the cost of such removal from the owners. Section 357 prohibits a person from depositing any rubbish etc., on the bank of a water course.

The pity is that at the level of implementation the law is in suspended animation for extraneous reasons. And the municipal body fails functionally because it does not provide sanitary facilities. The municipal body is often the environment offender!

The Merchant Shipping Act, 1958 deals with marine pollution enacted pursuant to an International Convention of 1954. It tackles a specific problem, and has a modern look. But the statute, par excellence, on pollution is the Water (Prevention and Control of Pollution) Act, 1974. Though water is a State subject, except that the Central Government has the responsibility of regulation and development of interstate rivers and valleys to the extent to which such regulation and development is declared by Parliament to be expedient in the public interest, yet the Act is a Central enactment. It has been enacted by Parliament under Art. 252 of the Constitution which enables the States to pass resolutions in their legislatures enabling Parliament to pass an Act for any matter falling in the State list. The Act applies to the States of Assam, Bihar, Gujarat, Haryana, Himachal Pradesh, Jammu and Kashmir, Karnataka, Kerala, Madhya Pradesh, Rajasthan, Tripura, Punjab, Uttar Pradesh, Andhra Pradesh, West Bengal and the Union Territories.

The Act is quite comprehensive in its coverage of various types of waters. It applies to streams which term has been defined to include rivers, water course (whether flowing or for the time being dry); inland water (whether natural or artificial); sub-terranean waters sea or tidal waters. The Act makes provisions for the establishment of a Central Board for Prevention and Control of Water Pollution and the State Boards for the purpose. These are autonomous boards.

In fact, Boards are created for enforcement. The two important function of the State Boards are; to lay down standards of pollution and to make consent orders for putting trade and sewage effluent into the streams. The Central Board is bound by the directions in writing as the Central Government may give. Every State Board is bound by the directions in writing as the Central Board or the State Government may give. Where a direction given by the State Governments is inconsistent with the direction by the Central Board, the matter is to be referred to the Central Government for its decision. There is a provision in the Act for appeal against the consent orders of the State Board to such an authority as the State Government may think fit to constitute. The State Government has also been given the power of revision over the orders of the Board. For contravening the standards laid down by the State Board and for violating the provisions relating to the consent by the Board, stiff penalties have been provided i.e. imprisonment which shall not be less than six months but which may tend to six years and fine. However, no court is to take cognisance of any offence under the Act except on a complaint made by, or with the previous sanction of, the State Board. The Act also contains provisions for the State Board to make application tq the court for restraining apprehended pollution of water. The Boards are given ample powers of investigation and inspection, and to take samples. There are provisions for establishing the Central Water Laboratory and the State Water Laboratory and appointment of analysts. Powers have been given to the State Boards to carry out themselves certain works relating to prevention and control of sewage or trade effluent.

The Act, as usual, is more or less a carbon copy of a like British legislation and has not made its presence felt by the community either of polluters or victims. Jurists have pointed out the defects in the Act and in some senses the departure from the British Act has made the Indian Act weaker. There is the problem of multiplicity of authorities, multiplicity of controls, multiplicity of enactments. Plurality promotes debility, not effectiveness. Moreover, the affected groups cannot come to the court directly without the sanction of the Board. This alienates the people from the law.

An untapped legal resource for control of pollution which has great potential is the law of torts. In India tort law is a neglected child while in America it is a pampered child. After Lord Atkin's magnificient expansion of the doctrine of negligence vis a vis injury to and duty towards neighbours in Donoghue V. Stevenson and even before that the Rule in Ryland Vs. Fletcher provided a sound basis for action against polluters. The law of nuisance also could be pressed into service. And if the dynamics of access jurisprudence enabling public interest litigation at the instance of those who are not directly injured were to be accepted by courts as settled law, the civil remedy by way of injunction or other specific relief and of damages could be a potent weapon, preventive, mandatory and otherwise. Why, the Indian law can build on and broaden the doctrine beyond Lord Atkin. Third World jurisprudence must enter fresh woods and pastures new even as English law did when economic developments demanded. We can and must make new strides in tort law to meet the challenges of environmental pollution. Deforestation even by Government, if the public will suffer irreversible trauma, must be a tort as an extention of public law. After all, torts are judicial creations. We need new-frontier men of the law in environmental protection. Legal Futurologists will agree that the environmental lawyer, with new orientation and skill plus a multi-disciplinary bias is the hope of the profession as avant garde servant of society. And then the judges; with creative vision, environmental perception and liberal concept of cause of action will declare pollution, exceeding safe levels, as a civil wrong to be stopped by quia timet action, mandatory injunction or damages for community injury, at the instance of people's pressure groups and pro bono litigators.

No man is free if his neighbour owes no duty not to injure him by polluting the air he breathes, the water he drinks, the land which gives him food and Nature which is his biosphere.

There is a social law of motion for judge-made law. For,

the law is not an ass but reacts when justice and freedom are threatened. And the court is functionally relevant in control of pollution because a tort, conceptually, is any injury even to the public, generally, once we remove from the eyes of the law the scales of individualism and substitute the lens of collectivism.

But the pathology of legal impotence in overcoming pollution is not so much that we have not enough law but that most-laws with police power bark but do not bite. The reason is four-fold. Environmental justice is entrusted to bureau-cratic machinery which, defunctiorially, has passion for files, not for people. Secondly, people's agencies, even associations of non-official experts, cannot directly take action and the Board is a bottleneck. Most importantly, the criminals have clout and so Board's bow, never beat. And, the law is genetically so clumsy that quick action is lawlessness.

What are the remedies for the maladies? A Pachasheel Marg may be considered.

1. Public interest litigation with broadened rules of locus standi for initiating or intervening in civil or criminal proceedings.
2. A Right to Information Act which entitles any interested person to seek and secure all, information bearing on any matter of public interest whether it be with a public body or, in special cases or good grounds, in the possession of private person's dealing with the subject.
3. Special Courts, should be created with sitting judges of the High Court, presiding over them, called Environment Protection Courts on the analogy of the Election Tribunals and Special Courts created

> for trial of V.I.Ps. These courts should be subject only to the Supreme Court. For this purpose, more judges should be appointed in the High Courts on the basis that expeditious disposal is of the highest moment. These judges must be free to devise their own procedure subject to basic rules of fair play. They may blend adversarial and inquisitorial methods and be released from the rigours of the Evidence Act.

Those fundamentalists of the common law who are upset by the procedural liberalism and judicial activism may read Dr. Zeidler who compared the two systems thus:

> Professor Kotz of Hamburg recently, at an annual conference of the German-British Jurists' Association, compared the two procedural systems by a parable: amongst the procedural systems the common law procedure is what a shining Rolls Royce car is amongst the automobiles, whereas the German procedure may be compared to a dusty small Volkswagen. I agree, but the question remains: what is it you can afford to pay for, and how often and in what situations are you in need of a Rolls Royce or of a Volkswagen? This leads me to those aspects of the common law procedure, which in my view may be looked upon as its handicaps. Under its rule most of the work is done by advocates. The parties have to pay for it accordingly. This means that people with education, financial resources and social and political connections have a better chance than the majority without such advantages. In Germany most of the legal work is done by judges, who get their salaries from the tax-payer, so there is no difference as regards the social or educational level as a starting point. Of course you may fall into the hands of a bad judge as well of bad advocates. But there is still a difference. The

> counsel you select yourself. The selection of the judge is a matter of destiny. So the chance of having a good judge is irrelevant to social status. It is the fair and equal chance of an honest gambling table.

The processual informality, the liberal locus standi, the powerful industrial adversary, the catalyst and activist role of the justicer and the ability to take affirmative action, interim and final, with multi-disciplinary perceptions and creative pragmatism - these call for a calibre not readily bought in the judicial market but carefully handpicked after intensive search for the right man. Enemies of Environment are corporate giants sometimes misguided Government itself as is alleged in the Silent Valley case or public sector enterprises, unconcerned with the nation's aesthetics, apart from economics, as was alleged in the Mathura Refinery and the Taj Mahal. To do justice between them and the meek, mobbish and not-so-well informed or well-financed community in opposition demands intelligent integrity and democratic sensitivity, detachment-cum-discernment, and juristic creativity, not the routine cup of tea of the robed fraternity, whatever some of them thinks of themselves with contempt power in their pockets.

It is apt to quote here Justice Khanna, a brave conservative and honest realist:

> 'No paper reform of the judicial system would prove effective unless such reform is implemented and translated into reality. For this purpose, we need the active co-operation of the members of the Bench and the Bar. It would also, in a great measure, depend upon the personality of the-judges. You may have hundred and one reforms but unless you have officers of the right calibre to preside over the courts, the reform would prove illusory and thus fail to deliver the goods. It is, therefore, essential to take every

care to ensure that the persons of the right calibre are selected to preside over the courts.

The fourth reform proposed is an Environment Ombudsman whose soul is beyond purchase or capable of command performer, whose spirit is restless at sight of social injustice, whose procuratorial zest reaches out to investigate and prosecute where legitimate grievances come into his ken. Indian experience in ombudsmanship is dubious. The Lokpal bill failed in Parliament with moralist Janata Party in power. However now it has been passed, leaving loops and holes. The mini-counterparts at State level, often superannating splendours, have functioned without hope for the aggrieved and worse without despair for the delinquent. We want Nature's advocate so that Spring may sing and not be silent, the Sun may not be shy to shine in smog, the woods may be lovely, green and deep, Gandhi may live a hundred years more and air, water and atmosphere may be pure and fresh and full of health. Aesthetics, conscience and harmonious blend of progress without pollution - these must guide him. Any private person or voluntary agency may alert him and he must move. He must have full investigative powers and access to files. He must advise and administer departments and industries, acquire know-how from professionals and keep touch with people's organisations and pressure groups. He must prosecute, if necessary appeal to the Supreme Court and lobby with Governments and legislators for action. He must be the nexus between Law and the People, Nature and Man, Government and Democracy.

The Ombudsman is the remedial promise of socialist patriotism. He is an institution with a plurality of research teams and contacts and investigative staff. There may be, as in some countries, more than one, as a collegium. Indeed, we need many because the task is heavy and varied. This body may have local offices and officers like the Central Bureau of Intelligence.

There is force in a similar suggestion made by Sally Thomas. The plea is put forward that:

> 'Once a genuine problem is discovered the Ombudsman sets the whole pollution control machinery in gear. An investigatory procedure is immediately begun. Negotiations in order to reach a compromise are undertaken and finally the problem is solved through compensation or through other alternative means.

If there has been a breach of normal code of conduct, the Ombudsman, in his role of Prosecutor, may impose a fine or penalisation in any other way. He can act suo moto and undertake an investigation where he has a reasonable apprehension of a threat of pollution.

Smaller units of Ombudsman should be appointed in every State. These units should be given a great deal of freedom, but not absolute freedom in the process of recommendation, compensation and redressals.

Since enforcing the law through action against polluters is expensive and time consuming, public participation is always preferred. Public participation allows interested citizens, public agencies and corporations to voice their interests and ideas. Therefore the scope of locus standi is very wide in environmental law.'

Finally, we advocate a fresh-, comprehensive Environment Protection Code up-dating, unifying the authorities, simplifying lexically, legally and structurally, the several Acts in force, classifying the topics in separate chapters, prescribing for Environment Boards, Environment

Courts, Environment Ombudsmen, providing for public interest litigation, affirmative action, wider rules of access to justice, combining, in one action, civil and criminal remedies and vesting the power to enforce orders. Provision must be made for appeals to the Special Environment Division of the Supreme Court.

Indeed, Indian Law, now, has no separate Noise Pollution Prevention Act. This is surely necessary and must be a chapter in the Code. The noise syndrome permeates every aspect of life. Even our politics, as the streets and legislatures testify, suffers, from noise sickness. Why our justicing - watch our courts - is combatively loquacious, their bench and the bar being contributories and the scene, with crowds and noises, is reminiscent of babel at times.

Our public life and its noises apart, here my concern, why, consternation, is about the injurious escalation of sounds from many sources which, if ignored, may inflict psychological, physiological and societal maladies. So we cannot be silent about the sound menace.

Noise has become a problem of importance to an industrial society as a result of the extensive use of machinery. Modern industries use machines, -vehicles and processes which consume large amount of energy. The human ear is constantly being assailed by man made sounds from all sides, and there remains few populous areas where relative quiet prevail.

Noise may have any one or all of the following effects on man. It may be changed a man's psychological state, may cause heart attack, cause chronic effects as hypertension or ulcers and damage his hearing. Further, noise not only damages hearing but also gives rise to many other ailments.

It is therefore, essential to control noise arising out of extensive use of machinery. In many countries there are legislations to control noise. England has a Noise Abatement Act; Japan has prescribed maximum limit for noise through legislation. Similarly, in several states in the U.S. law has prescribed maximum limit of noise. In India there is no specific legislation to control noise. However, the Workmen's Compensation Act provides for compensation for absolute deafness. The law shall not be silent on the issue of excessive sound which is actionable pollution.

Social action groups are the marines or spinal centres for activising the people and linking them with the legal process. Public interest litigation is the forensic manifestation of democratisation of judicial remedies, so necessary in the area of environmental law. Free legal aid, though conceptually different from public interest litigation, is a sister branch of socially sensitized jurisprudence. The need of the times is to expound a new philosophy of law and social philosophy, which will be theoretical foundation for the technology of free legal aid and public interest proceedings.

In developing countries, new departures from the individualism of industrialism are a legitimate demand of social democracy. Class actions, representative proceedings, organisations with special know-how coming to court in defence of the community - the list is long and takes many forms but postulates one fundamental. Justice is for the people collectively and can be sought by the people collectively. The men inevitability of public interest litigation flows from the imperative of social justice under the Constitution. The short issue is whether social justice is a judicial commitment, and, if yes, issues succulent with social justice or crying against social injustice, brought before the Court by people's collectives with Good Samaritan motives and without private incentives should be helped by overlooking procedural laxity and directing amicus curiae assistance or should be warned that procedure is the mistress, not the handmade, that judges open their jurisdiction to do justice only if forms are fully complied with what happens then is that courts of justice get estranged from the commoners in a country of poverty and law ceases to talk to justice. After all, even if the facts are not fully set out, courts must, once satisfied about the grievance and the handicap of the party, should not run away from its duty to do justice. Appoint an amicus curiae, issue directions to the State, collect the fuller facts, hear all sides and then take action.

> 'Our current processual jurisprudence is not of individualistic Anglo-Indian mould. It is broad-based and people-oriented and envisions access to justice through class actions, public Interest litigation, and representative proceedings. We have no hesitation in holding that the narrow concept of cause of action and person aggrieved and Individual litigation is becoming obsolescent in some jurisdiction. (AIR 1991 SC 298):

In the liberal states of the late 18th and 19th centuries the procedures for civil litigation reflected the individualistic philosophy of rights then prevalent; a right of access meant the 'aggrieved individual's formal right to institute or defend a claim. Access might be a natural right but, like other natural rights, it did not require state action for its protection; the rights existed prior to the state, it was enough-that the state did not allow infringement by other. In this field, as in others, the state remained passive with respect to the ability in practice, of a party to use his legal right. The law was mainly concerned with property and majestically assumed that all had property.

Such notions could not survive 20th Century popular democracy. Widespread literacy, popular education, improved communications and universal suffrage have made obsolete the old concept of a properties elite; property is more evenly distributed, social welfare programs alleviate individual distress. In legal terms the first fruit was provision of legal aid; a social welfare mechanism to allow all citizens, whatever their financial position, to enforce their private legal rights, to defend their personal liberty, status and property.

Legal Aid, in Cappelletti's term the first wave towards access, put them on a par with their wealther brethren but it did not affect the interests sought to be protected. The second wave Cappelletti defines as 'the reforms aimed at providing legal representation for diffuse interest especially in the areas of consumer and environmental protection. It is an expression of wider concern; better-educated citizens demand an increasing part in the running of their society and the decisions of their government. A more crowded affluent society has developed ideological causes; racial tolerance, civil rights, environmental and consumer protection. These interests may tough liberty or property but chiefly they are expressions of the values individuals wish their society to respect; the old proprietary interest is of less relevance. The new approach was recently summarized in New Zealand:

> . . . today it is unreal to suggest that a person looks to the law solely to protect his interest in a narrow sense. It is necessary to do no more than read the newspapers to see the breadth of the interests that today's citizen expects the law to protect - and he expects the court where necessary to provide that protection. He is interested in results, not procedural niceties.

Protection of environment found a place even in the manifestos of the political parties in 1980 at the time of 6th general elections. Thus the Congress I manifesto "pressing deep concern at the depletion of our forests and wildlife mentioned: 'Projects which bring economic benefits must be so planned as to preserve and enhance-our national wealth, our flora and fauna. The Lok Dal stated that it will pursue a vigorous policy of protecting the environment'. Similar policy statements were contained in the manifestos of other political parties like the Janta Party and the Indian, National Congress (U)".

The naive court, innocent of the burgeoning learning on ecology and technology of pollution control and the grave dangers of a moral industrialization and agricultural techniques, is an engine of injustice operating from an ivory tower. The law must keep the promises, as the great Holmes said. The robed judge must hold the sword and the scales, both together being symbolic of social justice.

This condition calls for legal entitlement to information from public sources, not only to legislators thro' interpellations but to all Indians public minded enough to demand it for public purposes.

Justice K.K. Mathew has said:

> 'A popular Government without popular information or the means of acquiring it is but a prologue to a farce or tragedy or perhaps both. A 'people who means to be their governors must arm themselves with power which knowledge gives", said Jarnes Madison. This is an echo in the sphere of governing of what Milton stated in universal and everlasting terms more than three hundred years ago. 'Give me the liberty to know and argue freely according to the dictates of conscience above all liberties'. Neither

> the right to elect periodically their representatives to the legislative bodies nor their acountability to the people at the end of their period of stewardship is of much avail unless the representatives are made responsive to public opinion when they decide public issues. This governing aspect of the people is expressed in the familiar statement that democracy has always rested upon the support of public opinion. The necessary condition of an enlightened public opinion is the existence of a well-informed citizenry'.

Justice Brandeis has said that the greatest menace to freedom is an inert people and that public discussion is a political duty in a republican form of Government. To fulfil this duty the people must have the necessary information, for they cannot carry on discussion in the void. It is this duty which generates the political right to know. The effective performance of the duty depends upon the extent to which the right has been actually realized. It is no doubt a right which the people are helpless to claim, for they do not know what they have a right to know what as yet they do riot know.

And the only means of acquiring information, without which there cannot be a well-informed citizenry is through freedom of a access to the materials in the possession of Government and also to proceedings of Government. The guarantee of freedom of speech includes the right to publish whatever secret information the press is able to obtain unless the act is reasonable. A free press in a democratic society postulates the right of access to all Governmental proceedings affecting the public, compatible only with the most convincing grounds of national security. Without this right of access, the freedom of the press is an empty guarantee.

Information is the bloodstream of democracy. To diminish information to people of the Government is to diminish their participation in Government. As Jefferson said, information to the people is the most certain and legitimate engine of Government. When a Government refuses to put its trust in the people, the people in turn will withdraw their trust from the Government. 'There is a strong public feeling against Secrecy of any kind in administration of Government', said President Polak. The secrecy system has become much less a means by which Government protects national security than a means by which Government safeguards its reputation, dissembles its purpose, buries its mistakes, manipulates its citizens, and maximizes its power and corrupts itself.

A rule requiring secrecy being established with disclosure being authorised only in exceptional cases, the path of the lowest resistance for bureaucracy is to follow the general rule and avoid the recognition of disconcerting exceptions. The bureaucratic concept of secrecy which we have inherited from the imperialistic tradition of the British administration still persists, though it marches ill with our democratic republic.

Woodrow Wilson expressed in an illuminating passage the necessity of publicity in all governmental matters. He said 'I for one have the conviction that Government ought to be all outside and not inside. I, for my part, believe that there ought to be no place where everything can be done that everyone does not know about. Everyone knows corruption thrives; in secret places and avoid public places'. We can lay it down as a fair presumption that secrecy means impropriety. Lord Acton had this in mind when he said nothing is safe that does not show that it can bear discussion and publicity. Proponents of the privilege themselves will admit that while the interest to be protected by with holding information may be important, still it will be outblanced by suspicious and adverse inferences which follow the claim of privilege.

In State of U.P Vs. Raj Narain, (1975) 4 SCC 428 the Supreme Court said:

> "In a Government of responsibility like ours, where all agents of the public must be responsible for their conduct there can be but few secrets. The people of this country have a right to know every public act, everything that is done in a public way by their public functionaries. They are entitled to know the particulars of every public transaction done in all its bearing. The right to know, which is derived from the concept of freedom of speech, though not absolute, is a factor which should make one wary when secrecy though not absolute, is a factor which is claimed for transactions which can at any rate have no repercussion on public security. To cover with the veil of secrecy the common routine business is not in the interest of the public. It is generally desired for purpose of parties and politics or personal self-interest or bureaucratic routine."

Before concluding let me emphasis again that legislation must envision positive action and not stop with policing the pollution. A national environmental policy and ecological plan must be promoted by law so that Man and Nature love and live in harmony.

The last pitched battle by man to save mankind against men hell-bent on homicidal environment maltreatment is programmed in the Namdapha Biosphere Reserve Report which I cite as an instance of the constructive direction which law must take to effectuate Art. 48A and 51A (g):

> "Art. 48A: Protection and improvement of environment and safeguarding of forests and wildlife - The State

> shall endeavour to protect and improve the environment and to safeguard the forests and wild life of the country."
>
> "Art. 51 A. (g): It shall be the duty of every citizen of India .... to protect and improve the natural environment including forests, lakes, rivers and wild life, and to have compassion for living creatures."

Remember, law must keep promises by reversing the self-destructive over-indulgence of commercial exploitation of Nature, burying up the tribal's innocence and raping virgin Earth, bleeding to-death Ibmorrow for To-day's profit and, in the process, blindly degrading eco-systems and destabilising the flow of progress. The Socialist Republic of India is a political-economic guarantee of social prosperity for generations to come and rejects the philosophy : Unborn tomorrow, dead yesterday,

Why fret then if to-day is sweat.

The future is not a beggar eating what left-overs the present leaves on the table but the rightful heir to the biqueathement of the estate preserved prudently as trustee.

The biosphere reserve project calls for legislation carefully drawn up with demarcation and protection of extensive areas where Nature may still sing and dance unmolested by Man. The legislative contemplation must include a Biosphere Reserve Government, Administration Council, Secretariat and monitoring the progress by Parliament on the basis of periodical reports. Many are the blessings of agricultural research, tourist potential, conservation education and model of science at the service of society - such a rich crop will wither away if law does not speak up now and forbid the spoilation and pollution busily working away currently in the devil's workshop. We have a Tomorrow if we fight the spectre of environmental counter-revolt-through Operation Biosphere Preservation. Man is his own prophet of doom, the liquidator of Nature's gift and, if he has the will, the architect of a progressive continuum for generations to come. This last is the socialist commitment of our Republic.

The biosphere sustains us all. To preserve it is the path to progress. To pollute it is the route to ruin-maybe to-day, maybe tomorrow. Survival through suicide is the nostrum of those who wish us to wink at pollution as the necessary price of progress. Humanism and law must call the bluff. To muddle and bungle, to omit and abet, the rape of the earth is to prove Bernard Shaw : Darwin taught us not that all men were monkeys once but that some monkeys have no tails now!

The Constitution commands us, the law forbids us. Let us be Natures patriots. Regulatory legislation to control environmental pollution is a must since the rule of law must defend the rule of life. And Life will survive only if the biosphere is safe. That is the integral yoga of Man and Nature. But, law is paper tiger unless education makes society militantly aware of the risks of pollution.

Appendices

APPENDIX—I

RIO DECLARATION ON ENVIRONMENT AND DEVELOPMENT

The United Nations Conference on Environment and Development,

Having met at Rio de Janeiro from 3 to 14 June 1992,

Reaffirming the Declaration of the United Nations Conference on the Human Environment, adopted at Stockholm on 16 June 1972, and seeking to build upon it,

With the goal of establishing a new and equitable global partnership through the creation of new levels of cooperation among States, key sectors of societies and people,

Working towards international agreements which respect the interests of all and protect the integrity of the global environmental and developmental system,

Recognising the integral and interdependent nature of the Earth, our home,

Proclaims that:

Principle 1

Human beings are at the centre of concerns for sustainable development. They are entitled to a healthy and productive life in harmony with nature.

Principle 2

States have, in accordance with the Charter of the United Nations and the principles of international law, the sovereign right to exploit their own resources pursuant to their own environmental and developmental policies, and the responsibility to ensure that activities within their jurisdiction or control do not cause damage to the environment of other States or of areas beyond the limits of national jurisdiction.

Priciple 3

The right to development must be fulfilled so as to equitably meet developmental and environmental needs of present and future generations.

Principle 4

In order to achieve sustainable development, environmental protection shall constitute an integral part of the development process and cannot be considered in isolation from it

Principle 5

All States and all people shall cooperate in the essential task of eradicating poverty as an indispensable requirement for sustainable development, in order to decrease the disparities in standards of living and better meet the needs of the majority of the people of the world.

Principle 6

The special situation and needs of developing countries, particularly the least developed and those most environmentally vulnerable, shall be given special priority. International actions in the field of environment and development should also address the interests and needs of all countries.

Principle 7

States shall cooperate in a spirit of global partnership to conserve, protect and restore the health and integrity of the Earth's ecosystem. In view of the different contributions to global environmental degradation, States have common but differentiated responsibilities. The developed countries acknowledge the responsibility that they bear in the international pursuit of sustainable development in view of the pressures their societies place on the global environment and of the technologies and financial resources they command.

Principle 8

To achieve sustainable development and a higher quality of life for all people, States should reduce and eliminate unsustainable patterns of production and consumption and promote appropriate demographic policies.

Principle 9

States should cooperate to strengthen endogenous capacity-building for sustainable development by improving scientific understanding through exchanges of scientific and technological knowledge, and by enhancing the development, adaptation, diffusion and transfer of technologies, including new and innovative technologies.

Principle 10

Environmental issues are best handled with the participation of all concerned citizens, at the relevant level. At the national level, each individual shall have appropriate access to information concerning the environment that is held by public authorities, including information on hazardous materials and activities in their communities, and the opportunity to participate in decision-making processes. States shall facilitate and encourage public awareness and participation by making information widely available. Effective access to judicial and administrative proceedings, including redress and remedy, shall be provided.

Principle 11

States shall enact effective environmental legislation. Environmental standards, management objectives and priorities should reflect the environmental and developmental context to which they apply. Standards applied by some countries may be inappropriate and of unwarranted economic and social cost to other countries, in particular developing countries.

Principle 12

States should cooperate to promote a supportive and open international economic system that would lead to economic growth and sustainable development in all countries, to better address the problems of environmental degradation. Trade policy measures for environmental purposes should not constitute a means of arbitrary or unjustifiable discrimination or a disguised restriction on international trade. Unilateral actions to deal with environmental challenges outside the jurisdiction of the importing country should be avoided. Environmental measures addressing transboundary or global environmental problems should, as far as possible, be based on an international consensus.

Principle 13

States shall develop national law regarding liability and compensation for the victims of pollution and other environmental damage. States shall also cooperate in an expeditious and more determined manner to develop further international law regarding liability and compensation for adverse effects of environmental damage caused by activities within their jurisdiction or control to areas beyond their jurisdiction.

Principle 14

States should effectively cooperate to discourage or prevent the relocation and transfer to other States of any activities and substances that cause severe environmental degradation or are found to be harmful to human health.

Principle 15

In order to protect the environment, the precautionary approach shall be widely applied by States according to their capabilities. Where there are threats of serious or irreversible damage, lack of full scientific certainty shall not be used as a reason for postponing cost-effective measures to prevent environmental degradation.

Principle 16

National authorities should endeavour to promote the internalisation of environmental costs and the use of economic instruments, taking into account the approach that the polluter should, in principle, bear the cost of pollution, with due regard to the public interest and without distorting international trade and investment.

Principle 17

Environmental impact assessment, as a national instrument, shall be undertaken for proposed activities that are likely to have a significant adverse impact on the environment and are subject to a decision of a competent national authority.

Principle 18

States shall immediately notify other States of any natural disasters or other emergencies that are likely to produce sudden harmful effects on the environment of those States. Every effort shall be made by the international community to help States so afflicted.

Principle 19

States shall provide prior and timely notification and relevant information to potentially affected States on activities that may have a significant adverse transboundary environmental effect and shall consult with those States at an early stage and in good faith.

Principle 20

Women have a vital role in environmental management and development. Their full participation is therefore essential to achieve sustainable development.

Principle 21

The creativity, ideals and courage of the youth of the world should be mobilised to forge a global partnership in order to achieve sustainable development and ensure a better future for all.

Principle 22

Indigenous people and their communities and other local communities have a vital role in environmental management and development because of their knowledge and traditional practices. States should recognise and duly support their identity, culture and interests and enable their effective participation in the achievement of sustainable development

Principle 23

The environment and natural resources of people under oppression, domination and occupation shall be protected.

Principle 24

Warfare is inherently destructive of sustainable development. States shall therefore respect international law providing protection for the environment in times of armed conflict and cooperate in its further development, as necessary.

Principle 25

Peace, development and environmental protection are interdependent and indivisible.

Principle 26

States shall resolve all their environmental disputes peacefully and by appropriate means in accordance with the Charter of the United Nations.

Principle 27

States and people shall cooperate in good faith and in a spirit of partnership in the fulfilment of the principles embodied in this Declaration and in the further development of international law in the field of sustainable development.

Non-Legally Binding Authoritative Statement of Principles for a Global Consensus on the Management, Conservation and Sustainable Development of All Types of Forests

PREAMBLE

(a) The subject of forests is related to the entire range of environmental and development issues and opportunities, including the right to socio-economic development on a sustainable basis.

(b) The guiding objective of these principles is to contribute to the management, conservation and sustainable development of forests and to provide for their multiple and complementary functions and uses.

(c) Forestry issues and opportunities should be examined in a holistic and balanced manner within the overall context of environment and development, taking into consideration the multiple functions and uses of forests, including traditional uses, and the likely economic and social stress when these uses are constrained or restricted, as well as the potential for development that sustainable forest management can offer.

(d) These principles reflect a first global consensus on forests. In committing themselves to the prompt implementation of these principles, countries also decide to keep them under assessment for their adequacy with regard to further international cooperation on forest issues.

(e) These principles should apply to all types of forests, both natural and planted, in all geographical regions and climatic zones, including austral, boreal, subtemperate, temperate, subtropical and tropical.

(f) All types of forests embody complex and unique ecological processes which are the basis for their present and potential capacity to provide resources to satisfy human needs as well as environmental values, and as such their sound management and conservation is of concern to the Governments of the countries to which they belong and are of value to local communities and to the environment as a whole.

(g) Forests are essential to economic development and the maintenance of all forms of life.

(h) Recognising that the responsibility for forest management, conservation and sustainable development is in many States allocated among federal/national, state/provincial and local levels of government, each State, in accordance with its constitution and/or national legislation, should pursue these principles at the appropriate level of government.

PRINCIPLES/ELEMENTS

1. (a) States have, in accordance with the Charter of the United Nations and the principles of international law, the sovereign right to exploit their own resources pursuant to their own environmental policies and have the responsibility to ensure that activities within their jurisdiction or control do not cause damage to the environment of other States or of areas beyond the limits of national jurisdiction.

   (b) The agreed full incremental cost of achieving benefits associated with forest conservation and sustainable development requires increased international cooperation and should be equitably shared by the international community.

2. (a) States have the sovereign and inalienable right to utilise, manage and develop their forests in accordance with their development needs and level of socio-economic development and on the basis of national policies consistent with sustainable development and legislation, including the conversion of such areas for other uses within the overall socio-economic development plan and based on rational land-use policies.

(b) Forest resources and forest lands should be sustainably managed to meet the social, economic, ecological, cultural and spiritual needs of present and future generations. These needs are for forest products and services, such as wood and wood products, water, food, fodder, medicine, fuel, shelter, employment, recreation, habitats for wildlife, landscape diversity, carbon sinks and reservoirs, and for other forest products. Appropriate measures should be taken to protect forests against harmful effects of pollution, including air-borne pollution, fires, pests and diseases, in order to maintain their full multiple value.

(c) The provision of timely, reliable and accurate information on forests and forest ecosystems is essential for public understanding and informed decision-making and should be ensured.

(d) Governments should promote and provide opportunities for the participation of interested parties, including local communities and indigenous people, industries, labour, non-governmental organisations and individuals, forest dwellers and women, in the development, implementation and planning of national forest policies.

3. (a) National policies and strategies should provide a framework for increased efforts, including the development and strengthening of institutions and programmes for the management, conservation and sustainable development of forests and forest lands.

(b) International institutional arrangements, building

on those organisations and mechanisms already in existence, as appropriate, should facilitate international cooperation in the field of forests.

(c) All aspects of environmental protection and social and economic development as they relate to forests and forest lands should be integrated and comprehensive.

4. The vital role of all types of forests in maintaining the ecological processes and balance at the local, national, regional and global levels through, *inter alia,* their role in protecting fragile ecosystems, watersheds and freshwater resources and as rich storehouses of biodiversity and biological resources and sources of genetic material for biotechnology products, as well as photosynthesis, should be recognised.

5. (a) National forest policies should recognise and duly support the identity, culture and the rights of indigenous people, their communities and other communities and forest dwellers. Appropriate conditions should be promoted for these groups to enable them to have an economic stake in forest use, perform economic activities, and achieve and maintain cultural identity and social organisation, as well as adequate levels of livelihood and well-being, through, *inter alia,* those land tenure arrangements which serve as incentives for the sustainable management of forests.

(b) The full participation of women in all aspects of the management, conservation and sustainable development of forests should be actively promoted.

6. (a) All types of forests play an important role in meeting energy requirements through the provision of a renewable source of bio-energy, particularly in developing countries, and the demands for fuelwood for household and industrial needs should be met through sustainable forest management, afforestation and reforestation. To this end, the potential contribution of plantations of both indigenous and introduced species for the provision of both fuel and industrial wood should be recognised.

(b) National policies and programmes should take into account the relationship, where it exists, between the conservation, management and sustainable development of forests and all aspects related to the production, consumption, recycling and/or final disposal of forest products.

(c) Decisions taken on the management, conservation and sustainable development of forest resources should benefit, to the extent practicable, from a comprehensive assessment of economic and non-economic values of forest goods and services and of the environmental costs and benefits. The development and improvement of methodologies for such evaluations should be promoted.

(d) The role of planted forests and permanent agricultural crops as sustainable and environmentally sound sources of renewable energy and industrial raw material should be recognised, enhanced and promoted. Their contribution to the maintenance of ecological processes, to offsetting pressure on primary/old-growth forest and to providing regional employment and development with the adequate involvement of local inhabitants should be recognised and enhanced.

(e) Natural forests also constitute a source of goods and services, and their conservation, sustainable management and use should be promoted.

7\. (a) Efforts should be made to promote a supportive international economic climate conducive to sustained and environmentally sound development of forests in all countries, which include, *inter alia,* the promotion of sustainable patterns of production and consumption, the eradication of poverty and the promotion of food security.

(b) Specific financial resources should be provided to developing countries with significant forest areas which establish programmes for the conservation of forests including protected natural forest areas. These resources should be directed notably to economic sectors which would stimulate economic and social substitution activities.

8\. (a) Efforts should be undertaken towards the greening of the world. All countries, notably developed countries, should take positive and transparent action towards reforestation, afforestation and forest conservation, as appropriate.

(b) Efforts to maintain and increase forest cover and forest productivity should be undertaken in ecologically, economically and socially sound ways through the rehabilitation, reforestation and re-establishment of trees and forests on unproductive, degraded and deforested lands, as well as through the management of existing forest resources.

(c) The implementation of national policies and programmes aimed at forest management, conservation and sustainable development, particularly in developing countries, should be supported by international financial and technical cooperation, including through the private sector, where appropriate.

(d) Sustainable forest management and use should be carried out in accordance with national development policies and priorities and on the basis of environmentally sound national guidelines. In the formulation of such guidelines, account should be taken, as appropriate and if applicable, of relevant internationally agreed methodologies and criteria.

(e) Forest management should be integrated with management of adjacent areas so as to maintain ecological balance and sustainable productivity.

(f) National policies and/or legislation aimed at management, conservation and sustainable development of forests should include the protection of ecologically viable representative or unique examples of forests, including primary/old-growth forests, cultural, spiritual, historical, religious and other unique and valued forests of national importance.

(g) Access to biological resources, including genetic material, shall be with due regard to the sovereign rights of the countries where the forests are located and to the sharing on mutually agreed terms of technology and profits from biotechnology products that are derived from these resources.

(h) National policies should ensure that environmental impact assessments should be carried out where actions are likely to have significant adverse impacts on important forest resources, and where such actions are subject to a decision of a competent national authority.

9. (a) The efforts of developing countries to strengthen the management, conservation and sustainable development of their forest resources should be supported by the international community, taking into account the importance of redressing external indebtedness, particularly where aggravated by the net transfer of resources to developed countries, as well as the problem of achieving at least the replacement value of forests through improved market access for forest products, especially processed products. In this respect, special attention should also be given to the countries undergoing the process of transition to market economies.

(b) The problems that hinder efforts to attain the conservation and sustainable use of forest resources and that stem from the lack of alternative options available to local communities, in particular the urban poor and poor rural populations who are economically and socially dependent on forests and forest resources, should be addressed by Governments and the international community.

(c) National policy formulation with respect to all types of forests should take account of the pressures and demands imposed on forest ecosystems and resources from influencing factors outside the forest sector, and intersectoral means of dealing with these pressures and demands should be sought.

10. New and additional financial resources should be provided to developing countries to enable them to sustainably manage, conserve and develop their forest resources, including through afforestation. reforestation and combating deforestation and forest and land degradation.

11. In order to enable, in particular, developing countries to enhance their endogenous capacity and to better manage, conserve and develop their forest resources, the access to and transfer of environmentally sound technologies and corresponding know-how on favourable terms, including on concessional and preferential terms, as mutually agreed, in accordance with the relevant provisions of Agenda 21, should be promoted, facilitated and financed, as appropriate.

12. (a) Scientific research, forest inventories and assessments carried out by national institutions which take into account, where relevant, biological, physical, social and economic variables, as well as technological development and its application in the field of sustainable forest management, conservation and development, should be strengthened through effective modalities, including international cooperation. In this context, attention should also be given to research and development of sustainably harvested non-wood products.

(b) National and, where appropriate, regional and international institutional capabilities in education, training, science, technology, economics, anthropology and social aspects of forests and forest management are essential to the conservation and sustainable development of forests and should be strengthened.

(c) International exchange of information on the results of forest and forest management research and development should be enhanced and broadened, as appropriate, making full use of education and training institutions, including those in the private sector.

(d) Appropriate indigenous capacity and local knowledge regarding the conservation and sustainable development of forests should, through institutional and financial support and in collaboration with the people in the local communities concerned, be recognised, respected, recorded, developed and, as appropriate, introduced in the implementation of programmes. Benefits arising from the utilisation of indigenous knowledge should therefore be equitably shared with such people.

13. (a) Trade in forest products should be based on non-discriminatory and multilaterally agreed rules and procedures consistent with international trade law and practices. In this context, open and free international trade in forest products should be facilitated.

(b) Reduction or removal of tariff barriers and impediments to the provision of better market access and better prices for higher value-added forest products and their local processing should be encouraged to enable producer countries to better conserve and manage their renewable forest resources.

(c) Incorporation of environmental costs and benefits into market forces and mechanisms, in order to achieve forest conservation and sustainable development, should be encouraged both domestically and internationally.

(d) Forest conservation and sustainable development policies should be integrated with economic, trade and other relevant policies.

(e) Fiscal, trade, industrial, transportation and other policies and practices that may lead to forest degradation should be avoided. Adequate policies, aimed at management, conservation and sustainable development of forests, including, where appropriate, incentives, should be encouraged.

14. Unilateral measures, incompatible with international obligations or agreements, to restrict and/or ban international trade in timber or other forest products should be removed or avoided, in order to attain long-term sustainable forest management.

15. Pollutants, particularly air-borne pollutants, including those responsible for acidic deposition, that are harmful to the health of forest ecosystems at the local, national, regional and global levels should be controlled.

APPENDIX—II

THE ENVIRONMENT (PROTECTION) ACT, 1986

MINISTRY OF ENVIRONMENT & FORESTS
GOVERNMENT OF INDIA

NO. 29 TO 1986

(23rd May, 1986)

An Act to provide for the protection and improvement of environment and for matters connected there with:

Whereas the decisions were taken at the United NationsConference on the Human Environment held at Stockholm in June, 1972, in which India participated, to take appropriate steps for the protection and improvement of human environment;

And whereas it is considered necessary further to implement the decisions aforesaid in so far as they relate to the protection and improvement of environment and the prevention of hazards to human beings, other living creatures, plants and property;

Be it enacted by Parliament in the Thirty-seventh Year of the Republic of India as follows:-

CHAPTER I

PRELIMINARY

1.Short Title, Extent and Commencement

(1) This Act may be Called the Environment (Protection) Act, 1986.

(2) It extends to the whole of India.

(3) It shall come into force on such date as the Central Government may, by notification in the Official Gazette, appoint and different dates may be appointed for different provisions of this Act and for different areas.

2. Definitions

In this Act, unless the context otherwise requires,—

(a) "environment" includes water, air and land and the inter-relationship which exists among and between water, air and land, and human beings, other living creatures, plants, micro-organism and property;

(b) "environmental pollutant" means any solid, liquid or gaseous substance present in such concentration as may be, or tend to be, injurious to environment;

(c) "environmental pollution" means the presence in the environment of any environmental pollutant;

(d) "handling", in relation to any substance, means the manufacture, processing, treatment, package, storage, transportation, use, collection, destruction, conversion, offering for sale, transfer or the like of such substance;

(e) "hazardous substance" means any substance or preparation which, by reason of its chemical or physico-chemical properties or handling, is liable to cause harm to human beings, other living creatures, plant, micro-organism, property or the environment;

(f) "occupier", in relation to any factory or premises, means a person who has, control over the affairs of the factory or the premises and includes in relation to any substance, the person in possession of the substance;

(g) "prescribed" means prescribed by rules made under this Act.

CHAPTER II

GENERAL POWERS OF THE CENTRAL GOVERNMENT

3. Power of Central Government to Take Measures to Protect and Improve Environment

(1) Subject to the provisions of this Act, the Central Government, shall have the power to take all such measures as it deems necessary or expedient

for the purpose of protecting and improving the quality of the environment and preventing controlling and abating environmental pollution.

(2) In particular, and without prejudice to the generality of the provisions of sub-section (1), such measures may include measures with respect to all or any of the following matters, namely:

(i) co-ordination of actions by the State Governments, officers and other authorities—

(a) under this Act, or the rules made thereunder, or

(b) under any other law for the time being in force which is relatable to the objects of this Act;

(ii) planning and execution of a nation-wide programme for the prevention, control and abatement of environmental pollution;

(iii) laying down standards for the quality of environment in its various aspects;

(iv) laying down standards for emission or discharge of environmental pollutants from various sources whatsoever:

Provided that different standards for emission or discharge may be laid down under this clause from different sources having regard to the quality or composition of the emission or discharge of environmental pollutants from such sources;

(v) restriction of areas in which any industries, operations or processes or class of industries, operations or processes shall not be carried out or shall be carried out subject to certain safeguards;

(vi) laying down procedures and safeguards for the prevention of accidents which may cause environmental pollution and remedial measures for such accidents;

(vii) laying down procedures and safeguards for the handling of hazardous substances;

(viii) examination of such manufacturing processes, materials and substances as are likely to cause environmental pollution;

(ix) carrying out and sponsoring investigations and research relating to problems of environmental pollution;

(x) inspection of any premises, plant, equipment, machinery, manufacturing or other processes, materials or substances and giving, by order, of such directions to such authorities, officers or persons as it may consider necessary to take steps for the prevention, control and abatement of environmental pollution;

(xi) establishment or recognition of environmental laboratories and institutes to carry out the functions entrusted to such environmental laboratories and institutes under this Act;

(xii) collection and dissemination of information in respect of matters relating to environmental pollution;

(xiii) preparation of manuals, codes or guides relating to the prevention, control and abatement of environmental pollution;

(xiv) such other matters as the Central

Government deems necessary or expedient for the purpose of securing the effective implementation of the provisions of this Act.

(3) The Central Government may, if it considers it necessary or expedient so to do for the purpose of this Act, by order, published in the Official Gazette, constitute an authority or authorities by such name or names as may be specified in the order for the purpose of exercising and performing such of the powers and functions (including the power to issue directions under section 5) of the Central Government under this Act and for taking measures with respect to such of the matters referred to in sub-section (2) as may be mentioned in the order and subject to the supervision and control of the Central Government and the provisions of such order, such authority or authorities may exercise and powers or perform the functions or take the measures so mentioned in the order as if such authority or authorities had been empowered by this Act to exercise those powers or perform those functions or take such measures.

4. Appontment of Officers and Their Powers and Functions

(1) Without prejudice to the provisions of sub-section (3) of section 3, the Central Government may appoint officers with such designation as it thinks fit for the purposes of this Act and may entrust to them such of the powers and functions under this Act as it may deem fit.

(2) The officers appointed under sub-section (1) shall be subject to the general control and direction of the Central Government or, if so directed by

that Government, also of the authority or authorities, if any, constituted under sub- section (3) of section 3 or of any other authority or officer.

5. Power to Give Directions

Notwithstanding anything contained in any other law but subject to the provisions of this Act, the Central Government may, in the exercise of its powers and performance of its functions under this Act, issue directions in writing to any person, officer or any authority and such person, officer or authority shall be bound to comply with such directions.

Explanation—For the avoidance of doubts, it is hereby declared that the power to issue directions under this section includes the power to direct—

(a) the closure, prohibition or regulation of any industry, operation or process; or

(b) stoppage or regulation of the supply of electricity or water or any other service.

6. Rules to Regulate Environment Pollution

(1) The Central Government may, by notification in the Official Gazette, make rules in respect of all or any of the matters referred to in section 3.

(2) In particular, and without prejudice to the generality of the foregoing power, such rules may provide for all or any of the following matters, namely:—

(a) the standards of quality of air, water or soil for various areas and purposes;

(b) the maximum allowable limits of concentration of various environmental pollutants (including noise) for different areas;

(c) the procedures and safeguards for the handling of hazardous substances;

(d) the prohibition and restrictions on the handling of hazardous substances in different areas;

(e) the prohibition and restriction on the location of industries and the carrying on process and operations in different areas;

(f) the procedures and safeguards for the prevention of accidents which may cause environmental pollution and for providing for remedial measures for such accidents.

CHAPTER III

PREVENTION, CONTROL, AND ABATEMENT OF ENVIRONMENTAL POLLUTION

7. Persons Carrying on Industry Operation, Etc., Not to Allow Emission or Discharge of Environmental Pollutants in Excess of the Standards

No person carrying on any industry, operation or process shall discharge or emit or permit to be discharged or emitted any environmental pollutants in excess of such standards as may be prescribed.

8. Persons Handling Hazardous Substances to Comply with Procedural Safeguards

No person shall handle or cause to be handled any hazardous substance except in accordance with such procedure and after complying with such safeguards as may be prescribed.

9. Furnishing of Information to Authorities and Agencies in Certain Cases

(1) Where the discharge of any environmental pollutant in excess of the prescribed standards occurs or is apprehended to occur due to any accident or other unforeseen act or event, the person responsible for such discharge and the person in charge of the place at which such discharge occurs or is apprehended to occur shall be bound to prevent or mitigate the environmental pollution caused as a result of such discharge and shall also forthwith—

(a) intimate the fact of such occurrence or apprehension of such occurrence; and

(b) be bound, if called upon, to render all assistance, to such authorities or agencies as may be prescribed.

(2) On receipt of information with respect to the fact or apprehension on any occurrence of the nature referred to in sub-section (1), whether through intimation under that sub-section or otherwise, the authorities or agencies referred to in sub-section (1) shall, as early as practicable, cause such remedial measures to be taken as necessary to prevent or mitigate the environmental pollution.

(3) The expenses, if any, incurred by any authority or agency with respect to the remedial measures referred to in sub-section (2), together with interest (at such reasonable rate as the Government may, by order, fix) from the date when a demand for the expenses is made until it is paid, may be recovered by such authority or agency from the person concerned as arrears of land revenue or of public demand.

10. Powers of Entry and Inspection

(1) Subject to the provisions of this section, any person empowered by the Central Government in this behalf shall have a right to enter, at all reasonable times with such assistance as he considers necessary, any place—

(a) for the purpose of performing any of the functions of the Central Government entrusted to him;

(b) for the purpose of determining whether and if so in what manner, any such functions are to be performed or whether any provisions of this Act or the rules made thereunder orany notice, order, direction or authorisation served, made, given or granted under this Act is being or has been complied with;

(c) for the purpose of examining and testing any equipment, industrial plant, record, register, document or any other material object or for conducting a search of any building in which he has reason to believe that an offence under this Act or the rules made thereunder has been or is being or is about to be committed and for seizing any such equipment, industrial plant, record, register, document or other material object if he has reason to believe that it may furnish evidence of the commission of an offence punishable under this Act or the rules made thereunder or that such seizure is necessary to prevent or mitigate environmental pollution.

(2) Every person carrying on any industry, operation or process of handling any hazardous substance shall be bound to render all assistance to the person empowered by the Central Government under sub-section (1) for carrying out the functions

under that sub-section and if he fails to do so without any reasonable cause or excuse, he shall be guilty of an offence under this Act.

(3) If any person wilfully delays or obstructs any persons empowered by the Central Government under sub-section (1) in the performance of his functions, he shall be guilty of an offence under this Act.

(4) The provisions of the Code of Criminal Procedure, 1973, or, in relation to the State of Jammu and Kashmir, or an area in which that Code is not in force, the provisions of any corresponding law in force in that State or area shall, so far as may be, apply to any search or seizures under this section as they apply to any search or seizure made under the authority of a warrant issued under section 94 of the said Code or as the case may be, under the corresponding provision of the said law.

11. Power to Take Simple and Procedure to be Followed in Connection Therewith

(1) The Central Government or any officer empowered by it in this behalf, shall have power to take, for the purpose of analysis, samples of air, water, soil or other substance from any factory, premises or other place in such manner as may be prescribed.

(2) The result of any analysis of a sample taken under sub-section (1) shall not be admissible in evidence in any legal proceeding unless the provisions of sub-sections (3) and (4) are complied with.

(3) Subject to the provisions of sub-section (4), the person taking the sample under sub-section (1) shall—

(a) serve on the occupier or his agent or person in charge of the place, a notice, then and there, in such form as may be prescribed, of his intention to have it so analysed;

(b) in the presence of the occupier of his agent or person, collect a sample for analysis;

(c) cause the sample to be placed in a container or containers which shall be marked and sealed and shall also be signed both by the person taking the sample and the occupier or his agent or person;

(d) send without delay, the container or the containers to the laboratory established or recognised by the Central Government under section 12.

(4) When a sample is taken for analysis under sub-section (1) and the person taking the sample serves on the occupier or his agent or person, a notice under clause (a) of sub-section (3), then,—

(a) in a case where the occupier, his agent or person wilfully absents himself, the person taking the sample shall collect the sample for analysis to be placed in a container or containers which shall be marked and sealed and shall also be signed by the person taking the sample, and

(b) in a case where the occupier or his agent or person present at the time of taking the sample refuses to sign the marked and sealed container or containers of the sample as required under clause (c) of sub-section (3), the marked and sealed container or containers shall be signed by the person taking the samples,

and the container or containers shall be sent without delay by the person taking the sample for analysis to the laboratory established or recognised under section 12 and such person shall inform the Government Analyst appointed or recognised under section 13 in writing, about the wilfull absence of the occupier or his agent or person, or, as the case may be, his refusal to sign the container or containers.

12.Environmental Laboratories

(1) The Central Government may, by notification in the Official Gazette,—

(a) establish one or more environmental laboratories;

(b) recognise one or more laboratories or institutes as environmental laboratories to carry out the functions entrusted to an environmental laboratory under this Act.

(2) The Central Government may, by notification in the Official Gazette, make rules specifying—

(a) the functions of the environmental laboratory;

(b) the procedure for the submission to the said laboratory of samples of air, water, soil or other substance for analysis or tests, the form of the laboratory report thereon and the fees payable for such report;

(c) such other matters as may be necessary or expedient to enable that laboratory to carry out its functions.

13. Government Analysis

The Central Government may by notification in the Official Gazette, appoint or recognise such persons as it thinks fit and having the prescribed qualifications to be Government Analysts for the purpose of analysis of samples of air, water, soil or other substance sent for analysis to any environmental laboratory established or recognised under sub-section (1) of section 12.

14. Reports of Government Analysts

Any document purporting to be a report signed by a Government analyst may be used as evidence of the facts stated therein in any proceeding under this Act.

15. Penalty for Contravention of the Provisions of the Act and Rules, Orders and Directions

(1) Whoever fails to comply with or contravenes any of the provisions of this Act, or the rules made or orders or directions issued thereunder, shall, in respect of each such failure or contravention, be punishable with imprisonment for a term which may extend to five years with fine which may extend to one lakh rupees, or with both, and in case the failure or contravention continues, with additional fine which may extend to five thousand rupees for every day during which such failure or contravention continues after the conviction for the first such failure or contravention.

(2) If the failure or contravention referred to in sub-section (1) continues beyond a period of one year after the date of conviction, the offender shall be punishable with imprisonment for a term which may extend to seven years.

16. Offences by Companies

(1) Where any offence under this Act has been committed by a company, every person who, at the time the offence was committed, was directly in charge of, and was responsible to, the company for the conduct of the business of the company, as well as the company, shall be deemed to be guilty of the offence and shall be liable to be proceeded against and punished accordingly:

Provided that nothing contained in this sub-section shall render any such person liable to any punishment provided in this Act, if he proves that the offence was committed without his knowledge or that he exercised all due diligence to prevent the commission of such offence.

(2) Notwithstanding anything contained in sub-section (1), where an offence under this Act has been committed by a company and it is proved that the offence has been committed with the consent or connivance of, or is attributable to any neglect on the part of, any director, manager, secretary or other officer of the company, such director, manager, secretary or other officer shall also deemed to be guilty of that offence and shall be liable to be proceeded against and punished accordingly.

Explanation—For the purpose of this section,—

(a) "company" means any body corporate and includes a firm or other association of individuals;

(b) "director", in relation to a firm, means a partner in the firm.

17. Offences by Government Department

(1) Where an offence under this Act has been committed by any Department of Government, the Head of the Department shall be deemed to be guilty of the offence and shall be liable to be proceeded against and punished accordingly:

Provided that nothing contained in this section shall render such Head of the Department liable to any punishment if he proves that the offence was committed without his knowledge or that he exercise all due diligence to prevent the commission of such offence.

(2) Notwithstanding anything contained in sub-section (1), where an offence under this Act has been committed by a Department of Government and it is proved that the offence has been committed with the consent or connivance of, or is attributable to any neglect on the part of, any officer, other than the Head of the Department, such officer shall also be deemed to be guilty of that offence and shall be liable to be proceeded against and punished accordingly.

CHAPTER IV

MISCELLANEOUS

18. Protection of Action Taken in Good Faith

No suit, prosecution or other legal proceeding shall lie against the Government or any officer or other employee of the Government or any authority constituted under this Act or any member, officer or other employee of such authority in respect of anything which is done or intended to be done in good faith in pursuance of this Act or the rules made or orders or directions issued thereunder.

19. Cognizance of Offences

No court shall take cognizance of any offence under this Act except on a complaint made by—

(a) the Central Government or any authority or officer authorised in this behalf by that Government, or

(b) any person who has given notice of not less than sixty days, in the manner prescribed, of the alleged offence and of his intention to make a complaint, to the Central Government or the authority or officer authorised as aforesaid.

20. Information, Reports or Returns

The Central Government may, in relation to its function under this Act, from time to time, require any person, officer, State Government or authority to furnish to it or any statistics, accounts and other information and such person, officer, State Government or other authority shall be bound to do so.

21. Members, Officers and Employees of the Authority Constituted Under Section 3 to be Public Servants

All the members of the authority, constituted, if any, in pursuance of any provisions of this Act or the rules made or orders or directions issued thereunder shall be deemed to be public servants within the meaning of section 21 of the Indian Penal Code (45 of 1860).

22. Bar of Jurisdiction

No civil court shall have jurisdiction to entertain any suit or proceeding in respect of anything done, action taken or order or direction issued by the Central Government or any other authority or officer in pursuance of any power conferred by or in relation to its or his functions under this Act.

23. Powers to Delegate

Without prejudice to the provisions of sub-section (3) of section 3, the Central Government may, by notification in the Official Gazette, delegate, subject to such conditions and limitations as may be specified in the notifications, such of its powers and functions under this Act [except the powers to constitute an authority under sub-section (3) of section 3 and to make rules under section 25] as it may deem necessary or expedient, to any officer, State Government or other authority.

24. Effect of Other Laws

(1) Subject to the provisions of sub-section (2), the provisions of this Act and the rules or orders made therein shall have effect notwithstanding anything inconsistent therewith contained in any enactment other than this Act.

(2) Where any act or omission constitutes an offence punishable under this Act and also under any other Act then the offender found guilty of such offence shall be liable to be punished under the other Act and not under this Act.

25. Power to Make Rules

(1) The Central Government may, by notification in the Official Gazette, make rules for carrying out the purposes of this Act.

(2) In particular, and without prejudice to the generality of the foregoing power, such rules may provide for all or any of the following matters, namely—

  (a) the standards in excess of which environmental pollutants shall not be discharged or emitted under section 7;

(b) the procedure in accordance with and the safeguards in compliance with which hazardous substances shall be handled or caused to be handled under section 8;

(c) the authorities or agencies to which intimation of the fact of occurrence or apprehension of occurrence of the discharge of any environmental pollutant in excess of the prescribed standards shall be given and to whom all assistance shall be bound to be rendered under sub-section (1) of section 9;

(d) the manner in which samples of air, water, soil or other substance for the purpose of analysis shall be taken under sub-section (1) of section 11;

(e) the form in which notice of intention to have a sample analysed shall be served under clause (a) of sub section (3) of section 11;

(f) the functions of the environmental laboratories, the procedure for the submission to such laboratories of samples of air, water, soil and other substances for analysis or test; the form of laboratory report; the fees payable for such report and other matters to enable such laboratories to carry out their functions under sub-section (2) of section 12;

(g) the qualifications of Government Analyst appointed or recognised for the purpose of analysis of samples of air, water, soil or other substances under section 13;

(h) the manner in which notice of the offence

and of the intention to make a complaint to the Central Government shall be given under clause (b) of section 19;

(i) the authority of officer to whom any reports, returns, statistics, accounts and other information shall be furnished under section 20;

(j) any other matter which is required to be, or may be, prescribed.

26. Rules Made Under This Act to be Laid Before Parliament

Every rule made under this Act shall be laid, as soon as may be after it is made, before each Hose of Parliament, while it is in session, for a total period of thirty days which may be ccmprised in one session or in two or more successive sessions, and if, before the expiry of the session immediately following the session or the successive sessions aforesaid, both Houses agree in making any modification in the rule or both Houses agree that the rule should not be made, the rule shall thereafter have effect only in such modified form or be of no effect, as the case may be; so, however, that any such modification or annulment shall be without prejudice to the validity of anything previously done under that rule.

Glossary

Abatement Debris: Waste from remediation activities.

Abatement: Reducing the degree or intensity of, or eliminating, pollution.

Ability to Treat Self and Others Fairly: To treat oneself and others equally, justly, and respectfully.

Activated Carbon: A highly adsorbent form of carbon used to remove odors and toxic substances from liquid or gaseous emissions. In waste treatment, it is used to remove dissolved organic matter from waste drinking water. It is also used in motor vehicle evaporative control systems.

Adverse Effects Data: FIFRA requires a pesticide registrant to submit data to EPA on any studies or other information regarding unreasonable adverse effects of a pesticide at any time after its registration.

Apathy (Opposite: Empathy): Lack of interest or concern; indifference.

Back Pressure: A pressure that can cause water to backflow into the water supply when a user's waste water system is at a higher pressure than the public system.

Bactericide: A pesticide used to control or destroy bacteria, typically in the home, schools, or hospitals.

Ballistic Separator: A machine that sorts organic from inorganic matter for composting.

Band Application: The spreading of chemicals over, or next to, each row of plants in a field.

Bar Screen: In wastewater treatment, a device used to remove large solids.

Bioassay: A test to determine te relative strength of a substance by comparing its effect on a test organism with that of a standard preparation.

Bioavailabiliity: Degree of ability to be absorbed and ready to interact in organism metabolism.

Blood Products: Any product derived from human blood, including but not limited to blood plasma, platelets, red or white corpuscles, and derived licensed products such as interferon.

Bloom: A proliferation of algae and/or higher aquatic plants in a body of water; often related to pollution, especially when pollutants accelerate growth.

Breakpoint Chlorination: Addition of chlorine to water until the chlorine demand has been satisfied.

Breakthrough: A crack or break in a filter bed that allows the passage of floc or particulate matter through a filter; will cause an increase in filter effluent turbidity.

Buffer Strips: Strips of grass or other erosion-resisting vegetation between or below cultivated strips or fields.

Building Cooling Load: The hourly amount of heat that must be removed from a building to maintain indoor comfort (measured in British thermal units (Btus).

Citing Evidence in Support of Ideas: To express, in oral or written form, information that confirms a particular view..

Civic Engagement: To participate in public life, encourage other people to participate in public life, and join in common work that promotes the well–being of everyone.

Civil and Political Rights –The rights to liberty and equality. Such rights include freedom to worship, to express oneself, to vote, to take part in political life, and to have access to information.

Classism: Attitude, action, and institutional practices that subordinate one class to a dominant class.

Creed: A set of fundamental beliefs or guiding principles. A creed can be either religious or secular.

Critical Thinking: Analyzing and contemplating past and present experiences, as well as future possibilities, by taking into account multiple perspectives on a story or narrative..

Critically Analyzing History: Analyzing a historical event or era from different cultural perspectives, including birth, gender, language, national or social origin, political or other opinion, property, or other status..

Cultural Rights: The right to preserve and develop one's cultural identity, as well as the right to protect endangered cultures.

Customary International Law: Law that becomes binding on nations through general acceptance as a matter of legal obligation. When enough states have begun to behave as though some principle is law, it becomes law "by use"; this is one of the main sources of international law.

Cynicism (Opposite: Optimism): An attitude of scornful or jaded negativity, especially a general distrust of the integrity or professed motives of others.

Deadmen: Anchors drilled or cemented into the ground to provide additional reactive mass for DP sampling rigs.

Decant: To draw off the upper layer of liquid after the heaviest material (a solid or another liquid) has settled.

Decay Products: Degraded radioactive materials, often referred to as "daughters" or "progeny"; radon decay products of most concern from a public health standpoint are polonium-214 and polonium-218.

Dechlorination: Removal of chlorine from a substance.

Decomposition: The breakdown of matter by bacteria and fungi

Democracy: A system of government in which people's views are reflected and the right of political participation is guaranteed. Such a form of government involves the principles of promoting and protecting human rights, social equality, and respect for the individual within a community.

Direct Action: Those tactics that can be undertaken by people themselves, without the help of government agencies, lawyers, or other institutions. Examples include picketing, work slowdowns, strikes, occupation of buildings, and marches.

Distinguishing Between Wants, Needs, and Rights: A *want* is a desire, whereas a *need* is a necessity and *rights* constitute entitlements. *Human rights* are mutual, equal, universal, non–discriminatory, inalienable, indivisible, and interdependent.

Diversity: The representation of multiple groups within a larger group, community, or area, such as a school or a workplace.

Ecological Entity: In ecological risk assessment, a general term referring to a species, a group of species, an ecosystem function or characteristic, or a specific habitat or biome.

Ecological Exposure: Exposure of a non-human organism to a stressor.

Ecological/Environmental Sustainability: Maintenance of ecosystem components and functions for future generations.

Ecosystem Structure: Attributes related to the instantaneous physical state of an ecosystem; examples include species population density, species richness or evenness, and standing crop biomass.

Ecosystem: The interacting system of a biological community and its non-living environmental surroundings.

Ecotone: A habitat created by the juxtaposition of distinctly different habitats; an edge habitat; or an ecological zone or boundary where two or more ecosystems meet.

Effluent Guidelines: Technical EPA documents which set effluent limitations for given industries and pollutants.

Effluent Limitation: Restrictions established by a state or EPA on quantities, rates, and concentrations in wastewater discharges.

Effluent: Wastewater—treated or untreated—that flows out of a treatment plant, sewer, or industrial outfall. Generally refers to wastes discharged into surface waters.

Emergency Removal Action: 1. Steps take to remove contaminated materials that pose imminent threats to local residents (e.g. removal of leaking drums or the excavation of explosive waste.) 2. The state record of such removals.

Emergency Response Values: Concentrations of chemicals, published by various groups, defining acceptable levels for short-term exposures in emergencies.

Emergency Suspension: Suspension of a pesticide product registration due to an imminent hazard. The action immediately halts distribution, sale, and sometimes actual use of the pesticide involved.

Empathy: The ability to listen deeply to another person's story or experience and connect to the person's feelings and story.

Energy Management System: A control system capable of monitoring environmental and system loads and adjusting HVAC operations accordingly in order to conserve energy while maintaining comfort.

Environmental Degradation: The diminishment of an ecosystem or a biosphere due to human activity. *Environmental degradation* occurs when natural resources (such as trees, habitat, earth, water, air) are consumed faster than they can be replenished. Sustainability requires that human activity, at a minimum, only uses nature's resources to the point where they can be replenished naturally.

Environmental, Cultural, and Developmental Rights: These rights recognize that people have the right to live in a safe and healthy environment and that groups of people have the right to cultural, political, and economic development.

Equality: This human rights principle mandates the same treatment of persons. The notion of fairness and respect for the inherent dignity of all human beings, as specified in Article 1 of the *Universal Declaration of Human Rights*: "All human beings are born free and equal in dignity and rights."

Ethnocentrism: A practice of consciously or unconsciously privileging one's own ethnic group over others that involves judging other groups by the values of one's own group.

Examining Assumptions: Process of deconstructing and analyzing the bases of ideas generally taken for granted in a given society or school of thought. An assumption is something taken for granted or accepted as true without proof, and can be a generalization based on limited or nonexistent experience. Identifying and considering the validity of assumptions are critical thinking skills.

Freedom: Political independence, *liberty*.

Genocide: A crime defined in international law as acts intended to destroy, in whole or in part, a national, ethnic, racial, or religious group of human beings.

Government: The act or process of governing, especially the decision-making and implementation of public policy in a political unit. The authority, leadership, or agent responsible for promoting and protecting human rights.

Government Responsibility: The obligations that a government must fulfill. Human Rights are an integral part of these obligations. Human rights are not gifts given at the pleasure of governments, nor should governments withhold them or apply them to some people but not to others. Governments must be held accountable for promoting and protecting the human rights of all persons.

Group Rights: The right of groups to protect their interests and identities.

Halon: Bromine-containing compounds with long atmospheric lifetimes whose breakdown in the stratosphere causes depletion of ozone. Halons are used in firefighting.

Hammer Mill: A high-speed machine that uses hammers and cutters to crush, grind, chip, or shred solid waste.

Hard Water: Alkaline water containing dissolved salts that interfere with some industrial processes and prevent soap from sudsing.

Hauler: Garbage collection company that offers complete refuse removal service; many will also collect recyclables.

Hazardous Substance: 1. Any material that poses a threat to human health and/or the environment. Typical hazardous substances are toxic, corrosive, ignitable, explosive, or chemically reactive. 2. Any substance designated by EPA to be reported if a designated quantity of the substance is spilled in the waters of the United States or is otherwise released into the environment.

Hazardous Waste Landfill: An excavated or engineered site where hazardous waste is deposited and covered.

Hazardous Waste Minimization: Reducing the amount of toxicity or waste produced by a facility via source reduction or environmentally sound recycling.

Hazardous Waste: By-products of society that can pose a substantial or potential hazard to human health or the environment when improperly managed. Possesses at least one of four characteristics (ignitability, corrosivity, reactivity, or toxicity), or appears on special EPA lists.

Human Rights: The rights people have simply because they are human beings, regardless of their ability, citizenship, ethnicity, gender, language, nationality, race, or sexuality; human rights become enforceable when they are *codified* as *conventions*, *covenants*, or *treaties*, as they become recognized as *customary international law*, or as they are accepted in national or local law.

Human Rights Commissions: Departments or committees of state and local governments in the United States established to address human rights issues by promoting human rights and investigating human rights violations.

Hurting People (feelings, physically): An action causing emotional, physical, psychological or spiritual pain, either intentionally or unintentionally, to another person or group.

Ignorance: The condition of being uneducated, unaware, or uninformed.

Illiteracy (Opposite: Literacy): The inability to read or write. Illiteracy can also refer to the ignorance of a set of terms or ideas that describe a concept. For example, human rights illiteracy is a lack of knowledge regarding human rights principles and norms.

Inalienable: Word that describes something that cannot be taken or given away. Human rights that individuals have cannot be taken away, surrendered, or transferred.

Individual Rights: A termreferring to what one is allowed to do and what can be done to an individual; the rights possessed by individuals.

Individuality vs. Collectivity: *Individuality* refers to the quality or state of being allowed to form one's own identity and being allowed to sustain that identity no matter what the identities of others are. *Collectivity* is a unification that is formed by more than one individual acting together.

Indivisible: Word that describes something that cannot be divided or reduced. Human rights should be addressed as an indivisible body, including civil, political, social, economic, cultural, and collective rights.

Injustice (Opposite: Justice): Denying fair, moral, and impartial treatment of the human rights of all persons.

Integrating Human Rights into Personal Awareness and Behaviors: To perceive one's own existence in terms of human rights, including one's own traits, feelings, and behaviors. Self–awareness is a personal understanding of the very core of one's own identity.

Interdependence: Human rights concerns appear in all spheres of life, such as in home, school, workplace, court, and markets. Human rights violations are interconnected; the loss of one right detracts from other rights. Similarly, promotion of human rights in one area supports other human rights.

International Labor Organization (ILO): Organization established in 1919 as part of the Versailles Peace Treaty to improve working conditions and promote social justice; the *ILO* became a specialized agency of the UN in 1946.

International Law: A set of rules generally regarded and accepted as binding in relations between states and nations. International human rights law is a part of international law designed to protect people against torture, inhuman treatment, arbitrary killings, discrimination, failure to take steps to provide adequate food, shelter, healthcare, and other human rights abuses.

Justice: Fairness, equity, and morality in action or attitude in order to promote and protect human rights and responsibilities. In most societies, people work for justice by organizing through different categories of rights, such as civil, political, economic, social, and cultural.

Kindness: Caring for or showing *empathy* for others.

Knowing Human Rights: To know the past, present, and evolving nature of human rights, as found in the *International Bill of Human Rights* and other human rights standards and norms.

Listening to and Understanding Diverse Perspectives: Exercising the ability to listen, understand, and apply diverse perspectives and stories to one's daily experiences.

Member States: Countries that are members of intergovernmental organizations, such as the United Nations.

Navigable Waters: Traditionally, waters sufficiently deep and wide for navigation by all, or specified vessels; such waters in the United States come under federal jurisdiction and are protected by certain provisions of the Clean Water Act.

Nitrate: A compound containing nitrogen that can exist in the atmosphere or as a dissolved gas in water and which can have harmful effects on humans and animals. Nitrates in water can cause severe illness in infants and domestic animals. A plant nutrient and inorganic fertilizer, nitrate is found in septic systems, animal feed lots, agricultural fertilizers, manure, industrial waste waters, sanitary landfills, and garbage dumps.

Nutrient Pollution: Contamination of water resources by excessive inputs of nutrients. In surface waters, excess algal production is a major concern.

Passivity: The condition of being inactive or submissive.

Prejudice: An attitude, opinion, or feeling formed without adequate prior knowledge, thought, or reason. *Prejudice* can be prejudgment for or against any individual, group, or object. Any individual or group can hold *prejudice(s)* towards another individual, group, or object.

Protocol: A *treaty* which modifies another *treaty* (e.g., adding additional procedures or substantive provisions).

Racism: An ideology of racial superiority and hierarchy based on discrimination.

Rule of Law: A government must function or operate in support of written laws, which should be adopted through an established procedure. This principle is intended to be a safeguard against unfair judgments or procedures in individual cases. Hence, those who make and enforce the law must respect and uphold the law.

Safe Water: Water that does not contain harmful bacteria, toxic materials, or chemicals, and is considered safe for drinking even if it may have taste, odor, color, and certain mineral problems.

Safe Yield: The annual amount of water that can be taken from a source of supply over a period of years without depleting that source beyond its ability to be replenished naturally in "wet years."

Safe: Condition of exposure under which there is a practical certainty that no harm will result to exposed individuals.

Saturated Zone: The area below the water table where all open spaces are filled with water under pressure equal to or greater than that of the atmosphere.

Yard Waste: The part of solid waste composed of grass clippings, leaves, twigs, branches, and other garden refuse.

Bibliography

Adams, Simon; David Lambert (2006). *Earth Science: An illustrated guide to science*. New York: Chelsea House.

Amnesty International (1995) *Human Rights Are Women's Right*. New York: Amnesty International.

Anaya, S. James (1996) *Indigenous Peoples in International Law*. Oxford: Oxford University Press.

An-Naim, Abdullahi Ahmed and Francis M. Deng, eds. (1990) *Human Rights in Africa: Cross-Cultural Perspectives*. Washington, D.C.: Brookings Institution.

An-Naim, Abdullahi Ahmed, ed. (1992) *Human Rights in Cross-Cultural Perspectives: A Quest for Consensus*. Philadelphia: University of Pennsylvania Press.

Arat, Zehra F. (1991) *Democracy and Human Rights in Developing Countries*. Boulder, Colorado: Lynne Rienner Publishers.

Begum, R. (1993) Women in Environmental Disasters: The 1991 Cyclone in Bangladesh. *Focus on Gender* I (1):34-39.

Braidotti, Rosi, Ewa Charkiewicz, Sabine Hausler, Saskia Wieringa (1993) *Women, the Environment, and Sustainable Development: Towards a Theoretical Synthesis*. London: Zed Books.

Brems, Eva (2001) *Human Rights: Universality and Diversity*. The Hague: Martinus Nijhoff Publishers.

Bunch, Charlotte (1990) Women's Rights as Human Rights:

Toward a Re-vision of Human Rights. *Human Rights Quarterly* 12(4):486-498.

Cassese, Antonio (1991) *Human Rights in a Changing World*. Philadelphia: Temple University Press.

Charlton, Sue Ellen M. (1984) *Women in Third World Development*. Boulder, Colorado: Westview Press.

Claude, Richard Pierre and Burns H. Weston, eds. ((1992) *Human Rights in the World Community: Issues and Action*. Second Edition. Philadelphia: University of Pennsylvania Press.

Domeisen, Natalie (1997) The Role of Women in Protecting Communities from Disasters. *Natural Hazards Observer* 21(5):5-6.

El Dareer, Asma (1982) *Woman, Why Do You Weep? Circumcision and Its Consequences*. London: Zed Press.

Enarson, Elaine and Betty Hearn Morrow, eds. (1999) *The Gendered Terrain of Disaster: Through Women's Eyes*. Westport, Connecticut: Greenwood Publishing.

Flynn, James, Paul Slovic, and C.K. Mertz (1994) Gender, Race, and Perception of Environmental Health Risks. *Risk Analysis* 14(6):1101-1108.

Forbes Martin, S. and J.Mends-Cole (1992) Refugee Women and Economic Self-Reliance. *Small Enterprise Development* 3(3).

Fordham, Maureen and Anne-Michelle Ketteridge (1996) Men Must Work and Women Must Weep: Examining Gender Stereotypes in Disaster. Paper presented at Natural Hazards Research and Applications Workshop, Denver, Colorado.

Forsythe, David P. (1994) *Human Rights in the New Europe: Problems and Progress*. Lincoln: University of Nebraska Press.

Forsythe, David P. (2000) *Human Rights in International Relations*. Cambridge: Cambridge University Press.

Fothergill, Alice (1996) Gender, Risk and Disaster. *International Journal of Mass Emergencies and Disasters* 14(1):33-56.

Fraser, Arvonne S. (1999) Becoming Human: The Origins and Development of Women's Human Rights. *Human Rights Quarterly* 21:853-906.

Freedman, Lynn (1995). Reflections on Emerging Frameworks of Health and Human Rights. *Health and Human Rights* 314, 333.

Fuller, Helene (1994) Development of Women's Policies for Emergencies and Disasters. *Australian Journal of Disaster Management* 9(2): 24-27.

Gray, Leslie (1993) The Effects of Drought and Economic Decline on Rural Women in Western Sudan. *Geoforum* 24(1):89-98.

Honeycombe, Beth (1994) Special Needs of Women in Emergency Situations. *Australian Journal of Emergency Management* 8(4):28-31.

Ikeda, Keiko (1995) Gender Differences in Human Loss and Vulnerability in Natural Disasters: A Case Study from Bangladesh. *Indian Journal of Gender Studies* 2 (2):171-193.

Jackson, Cecile (1993) Doing What Comes Naturally? Women and Environment in Development. *World Development* 21(12):1947-1963.

James, Valentide Udo, ed. (1995) *Women and Sustainable Development in Africa*. Westport, Econnecticut: Praeger Publishers.

Kafi, Sharif (1992) *Disasters and Destitute Women: Twelve Case Studies*. Dhaka: Bangladesh Development Partnership Centre.

Kassindja, Fauziya and Layil Miller Bashir (1998) *Do They Hear You When You Cry?* New York, New York: Delacorte Press.

Macklin, Audrey (1995) Refugee Women and the Imperative of Categories. *Human Rights Quarterly* 17(2):213-277.

Madzokere, C. (1993) Gender and Work - Past, Present and Future: The Situation of Rural Mozambican Women at Mazowe River Bridge Camp in Zimbabwe. *Journal of Social Development in Africa* 8(2):23-32.

Maluwa, Tiyanjana (1999) *Implementing the Principle of Gender Equality Through the Law: Some Lessons from Southern Africa.* Lincoln, Nebraska: International Human Rights and Human Diversity Initiative.

Mayer, Ann E. (1999) *Islam and Human Rights: Tradition and Politics.* Oxford and Boulder: Westview Press.

Milne, A.J.M. (1986) *Human Rights and Human Diversity: An Essay in the Philosophy of Human Rights.* Albany, New York: State University of New York Press.

Moser, Carolyn O.N. (1989) Gender Planning in the third World: Meeting Practical and Strategic Gender Needs. *World Development* 17(11):1799-1825.

Myers, Mary (1994) Women and Children First: Introducing a Gender Strategy Into Disaster Preparedness. *Focus on Gender* 2 (1): 14-16.

Neal, David, and Brenda Phillips (1990) Female-Dominated Local Social Movement Organizations in Disaster-Threat Situations. In *Women and Social Protest.* G. West and R. Blumberg, eds. Pp. 243-255. New York: Oxford University Press.

O'Hare, Ursala A. (1999) Realizing Human Rights for Women. *Human Rights Quarterly* 21(2):364-402.

Overseas Development Institute (1995) *Making Aid Work For Women*. London. ODI.

Peterson, Kristina (1997) From the Field: Gender Issues in Disaster Response and Recovery. *Natural Hazards Observer* 21(5):3-4.

Risse, Thomas, Stephen C. Ropp, and Kathryn Sikkink, eds. (1999) *The Power of Human Rights: International Norms and Domestic Change*. Cambridge: Cambridge University Press.

Shepherd, George W., Jr. and Ved P. Nanda, eds. (1985) *Human Rights and Third World Development*. Westport, Connecticut: Greenwood Press.

Smith, Gary A.; Aurora Pun (2006). *How Does the Earth Work?*. Upper Saddle River, NJ 07458: Pearson Prentice Hall. pp. 5.

Spinner-Halev, Jeff (1994) *The Boundaries of Citizenship: Race, Ethnicity, and Nationality in the Liberal State*. Baltimore: Johns Hopkins University Press.

Stalker, Peter, ed. (1995) *States of Disarray: The Social Effects of Globalization*. London: United Nations Research Institute for Social Development.

Index

K

L

M

N

O

P

R

S

■■■